MECHANICS OF FINITE DEFORMATION AND FRACTURE

MECHANICS OF FINITE DEFORMATION AND FRACTURE

Majid Aleyaasin
School of Engineering, University of Aberdeen,
Aberdeen, Scotland, UK

AAP | APPLE ACADEMIC PRESS

Apple Academic Press Inc.
3333 Mistwell Crescent
Oakville, ON L6L 0A2
Canada

Apple Academic Press Inc.
9 Spinnaker Way
Waretown, NJ 08758
USA

Exclusive worldwide distribution by CRC Press, a member of Taylor & Francis Group

No claim to original U.S. Government works

Printed in the United States of America on acid-free paper

International Standard Book Number-13: 978-1-77188-098-5 (Hardcover)

International Standard Book Number-13: 978-1-4987-1702-1 (eBook)

Typeset by Accent Premedia Services (www.accentpremedia.com).

Library and Archives Canada Cataloguing in Publication

Aleyaasin, Majid, author
Mechanics of finite deformation and fracture / Majid Aleyaasin (School of Engineering, University of Aberdeen, Aberdeen, Scotland, UK).

Includes bibliographical references and index.
Issued in print and electronic formats.
ISBN 978-1-77188-098-5 (hardcover).--ISBN 978-1-4987-1702-1 (pdf)
1. Deformations (Mechanics)--Mathematical models. 2. Fracture mechanics--Mathematical models. I. Title.

TA417.6.A44 2015 620.1'123 C2015-905606-3 C2015-905607-1

Library of Congress Cataloging-in-Publication Data

Aleyaasin, Majid.
Mechanics of finite deformation and fracture / Majid Aleyaasin.

pages cm
Includes bibliographical references and index.
ISBN 978-1-77188-098-5 (alk. paper)
1. Deformations (Mechanics) 2. Fracture mechanics. 3. Strains and stresses. I. Title.

TA417.6.A44 2015 620.1'123--dc23 2015030710

Dedicated to

my son, Sina,

and my daughter, Narges

CONTENTS

CONTENTS

ABOUT THE AUTHOR

Dr. Majid Aleyaasin, PhD

Majid Aleyaasin, PhD, is currently a researcher and lecturer in the School of Engineering at the University of Aberdeen in Aberdeen, Scotland. He was formerly a lecturer in mechanical engineering at Mashhad University, Iran, where he received a BEng. He received his PhD in mechanical engineering at Bradford University in the United Kingdom and subsequently worked as a research fellow at the University of Manchester Institute of Science and Technology and the University of Manchester, United Kingdom. Dr. Aleyaasin's research interest lies in the field of applied dynamics of solids and structures, and he has published 30 papers in international journals and in conference proceedings.

LIST OF ABBREVIATIONS

FEM	finite element method
SIF	stress intensity factor
EPFM	elastic-plastic fracture mechanics
LEFM	linear-elastic fracture mechanic
CTOD	crack tip opening displacement

PREFACE

Finite deformations in material can occur with change of a geometry, such that the deformed shape may not resemble the initial shape. Analyzing these types of deformations needs particular a mathematical tool, which is always associated with tensor notations. In general the geometry may be non-orthogonal and we need to use covariant and contra-variant tensor concepts to express the finite deformations and the associated mechanical strains.

Moreover, it is obvious that in large deformations, there are several definitions for stress; each depends on the frame of the stress definitions. The constitutive equations in material also depends on the type of stress, which is introduced. In simulation of the material deformation, components of the deformation tensor will be transformed from one frame to another either in orthogonal or in non-orthogonal coordinate of geometry.

Part one of this book contains 16 sections that are arranged in five chapters (1–5) and discusses the above issues in detail. There are some exercises in each chapter to help the reader to accomplish the derivation of the formulas. Once all 16 sections studied, the reader can fully understand the key concepts in mechanics of finite deformation. For more details concerning particular types of deformation the reader may seek further information from the valuable references listed in the end of part one of the book. Having acquired the knowledge in part one enables the reader to understand the advanced books and research papers in this field very easily.

Part two of the book contains chapters 6-10 and can be studied independently, without any knowledge about part one and covers the theory of fracture in brittle and quasi-brittle material. Quasi-brittleness is referred to the situation in which plastic deformation is not large when fracture occurs. This means nonlinear fracture mechanics in not covered in this part of the book. However, the linear-elastic fracture mechanics and also elastic-plastic fracture mechanics are fully explained in this part of the book.

Part two of this book contains 20 sections that are arranged in five chapters (6–10) and which studies the mechanics of fracture. As far as the author is aware, these chapters are the only source for a reader by which he/she can understand the key concepts in "fracture mechanics" without looking at other references (or books). Frankly this section is almost a book in itself on fracture mechanics, along with formulas and concepts. Like part one, there are also some exercises in each chapter, helping the reader to accomplish the derivation of the formulas.

For understanding part one of the book, the mathematical tools like metric tensors, etc., are necessary and are fully explained in the chapters. The only prerequisite is a knowledge of vectors calculus. For understanding part two of the book an elementary knowledge of complex variable theory is required. The advanced mathematical tools like conformal mapping and boundary integral method of Muskhelishvili are fully explained in the chapters of part two.

The graduate students in mechanical, civil, and material engineering will find this book useful in understanding the key concepts in material deformation and fracture. Part one of book can be thought as part of an advanced course in "nonlinear solid mechanics" for graduate students. Meanwhile part two of the book can be taught independently without any reference to part one as part of a graduate course in "fracture mechanics." This book is also useful for engineers and researchers in the field of solid and applied mechanics and will help them to understand the mathematically orientated material regarding finite deformation and fracture.

Majid Aleyaasin
School of Engineering, Aberdeen University,
Aberdeen Scotland, UK
July 2015

PART 1

MECHANICS OF FINITE DEFORMATIONS

CHAPTER 1

NONLINEAR GEOMETRY OF CONTINUUM SOLIDS VIA MATHEMATICAL TOOLS

CONTENTS

1.1 GENERAL NON-ORTHOGONAL SYSTEM OF COORDINATES

So far we are familiar with orthogonal coordinate systems. In solid mechanics sometimes we call it non-natural type. This is because a solid object may not be orthogonal in shape, and therefore applying orthogonal coordinate to study the mechanics of it naturally does not coincide with the geometry of the object.

According to Fig. 1.1, for the orthogonal coordinates the coordinates are X_1, X_2 and X_3 with the unit vectors $\vec{I}_1$, $\vec{I}_2$ and $\vec{I}_3$ that are fixed in space or are x_1, x_2 and x_3 with the unit vectors $\vec{i}_1$, $\vec{i}_2$ and $\vec{i}_3$ that are attached to the body.

In orthogonal system any vector $\vec{r}$ can be represented by:

$$\vec{r} = x_1\,\vec{i}_1 + x_2\,\vec{i}_2 + x_3\,\vec{i}_3 \quad \text{or} \quad \vec{r} = X_1\,\vec{I}_1 + X_2\,\vec{I}_2 + X_3\,\vec{I}_3$$

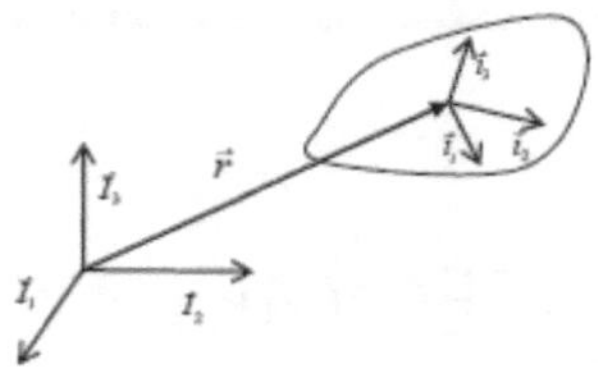

FIGURE 1.1 Orthogonal coordinate system.

For the unit vectors in orthogonal coordinate we have:

$$\vec{I}_1 \cdot \vec{I}_2 = \vec{I}_2 \cdot \vec{I}_3 = \vec{I}_3 \cdot \vec{I}_1 = 0 \qquad \vec{I}_1 \cdot \vec{I}_1 = \vec{I}_2 \cdot \vec{I}_2 = \vec{I}_3 \cdot \vec{I}_3 = 1$$

$$\vec{i}_1 \cdot \vec{i}_2 = \vec{i}_2 \cdot \vec{i}_3 = \vec{i}_3 \cdot \vec{i}_1 = 0 \qquad \vec{i}_1 \cdot \vec{i}_1 = \vec{i}_2 \cdot \vec{i}_2 = \vec{i}_3 \cdot \vec{i}_3 = 1$$

An important relationship between the unit vectors and the components in orthogonal system is:

$$x_1 = \vec{r} \cdot \vec{i}_1 \qquad x_2 = \vec{r} \cdot \vec{i}_2 \qquad x_3 = \vec{r} \cdot \vec{i}_3$$

$$X_1 = \vec{r} \cdot \vec{I}_1 \qquad X_2 = \vec{r} \cdot \vec{I}_2 \qquad X_3 = \vec{r} \cdot \vec{I}_3$$

It should be remembered that sign "·" is for internal vector product and sign "×" is for external vector product. This means that value of each component is the same as value of the projection of the vector into that axis. For some geometries like cylindrical, spherical, rectangular system of coordinate the above property is valid.

However, for many objects their natural geometry may be different for orthogonal ones, and a natural coordinates should be defined for them. For non-orthogonal coordinate we define the surfaces $\alpha_1 = c_1$, $\alpha_2 = c_2$ and $\alpha_3 = c_3$ to represent any point on the object and any position vector $\vec{r}$ according to Fig. 1.2 will be expressed by these surfaces for each point on the object three vectors $\vec{g}_1$, $\vec{g}_2$ and $\vec{g}_3$ can be defined to be replaced by the unit vectors $\vec{i}_1$, $\vec{i}_2$ and $\vec{i}_3$. These new vectors are defined by:

$$\vec{g}_1 = \frac{\partial \vec{r}}{\partial \alpha_1} \qquad \vec{g}_2 = \frac{\partial \vec{r}}{\partial \alpha_2} \qquad \vec{g}_3 = \frac{\partial \vec{r}}{\partial \alpha_3}$$

These vectors are called ($\vec{g}_1$, $\vec{g}_2$ and $\vec{g}_3$) "covariant based vectors." Obviously they may not be unit vectors and also they may not be orthogonal to each

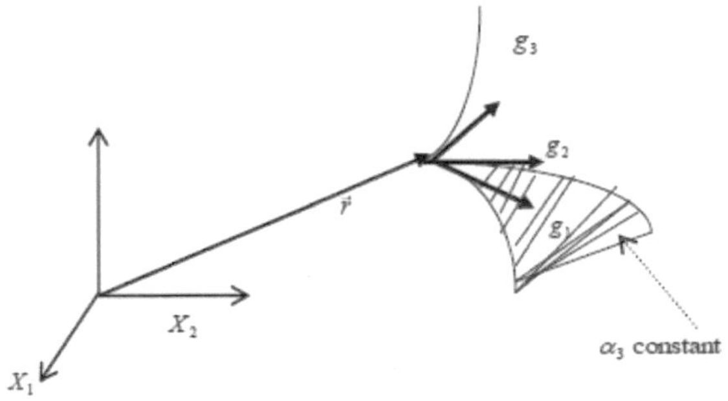

FIGURE 1.2 Non-orthogonal coordinate system.

other. $\vec{g}_1$, $\vec{g}_2$ and $\vec{g}_3$ originated from natural geometry of the object and any vector $\vec{r}$ can be represented with three components across $\vec{g}_1$, $\vec{g}_2$ and $\vec{g}_3$ respectively, that is,

$$\vec{r} = x^1\,\vec{g}_1 + x^2\,\vec{g}_2 + x^3\,\vec{g}_3 \tag{1.1.1}$$

The components x^1, x^2 and x^3 designated by upper subscripts will be explained shortly. The relationship between x^1, x^2 and x^3 with $\vec{g}_1$, $\vec{g}_2$ and $\vec{g}_3$ (they are not unit vectors) are not similar to orthogonal case, that is,

$$x^1 \neq (\vec{r}\cdot\vec{g}_1) \qquad x^2 \neq (\vec{r}\cdot\vec{g}_2) \qquad x^3 \neq (\vec{r}\cdot\vec{g}_3)$$

Obviously, value of each components x^1, x^2 and x^3 are not the same as value of the projections of the vector $\vec{r}$ into $\vec{g}_1$, $\vec{g}_2$ and $\vec{g}_3$, which are not unit vectors and form a parallelepiped as can be seen in Fig. 1.3. The volume of this parallelepiped is not unit but can be given by:

$$v = \vec{g}_1\cdot(\vec{g}_2\times\vec{g}_3) = \vec{g}_2\cdot(\vec{g}_1\times\vec{g}_3) = \vec{g}_3\cdot(\vec{g}_1\times\vec{g}_2) \tag{1.1.2}$$

Now an important question arises. If x^1, x^2 and x^3 are not the projections of vector $\vec{r}$ into $\vec{g}_1$, $\vec{g}_2$ and $\vec{g}_3$ then what are they? Can we find a direction $\vec{g}^i$ such that the projection of vector $\vec{r}$ into it is x^i? If the answer is yes then we call the system $\vec{g}^1$, $\vec{g}^2$ and $\vec{g}^3$ a **Contravariant base vector system. Einstein defined this vector base system by:**

$$\vec{g}^1 = \frac{\vec{g}_2\times\vec{g}_3}{v} \tag{1.1.3}$$

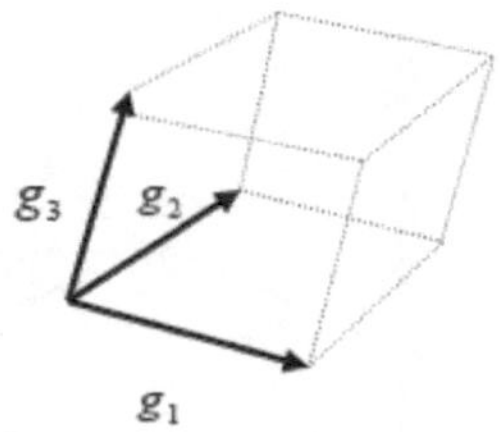

FIGURE 1.3 Parallelepiped of covariant base vectors.

$$\vec{g}^2 = \frac{\vec{g}_3 \times \vec{g}_1}{v} \tag{1.1.4}$$

$$\vec{g}^3 = \frac{\vec{g}_1 \times \vec{g}_2}{v} \tag{1.1.5}$$

From Eq. (1.1.3) we see that $\vec{g}^1$ is normal to the plane that is formed by $\vec{g}_2$ and $\vec{g}_3$ which means that $\vec{g}^1 \cdot \vec{g}_2 = 0$ and $\vec{g}^1 \cdot \vec{g}_3 = 0$. From Eq. (1.1.4) we see that $\vec{g}^2$ is normal to the plane that is formed by $\vec{g}_1$ and $\vec{g}_3$ which means that $\vec{g}^2 \cdot \vec{g}_1 = 0$ and $\vec{g}^2 \cdot \vec{g}_3 = 0$. From Eq. (1.1.5) we see that $\vec{g}^3$ is normal to the plane that is formed by $\vec{g}_1$ and $\vec{g}_2$ which means that $\vec{g}^3 \cdot \vec{g}_1 = 0$ and $\vec{g}^3 \cdot \vec{g}_2 = 0$.

Multiplying both sides of Eq. (1.1.3) into $\vec{g}_1$ considering Eq. (1.1.2) provides:

$$\vec{g}_1 \cdot \vec{g}^1 = \frac{\vec{g}_1 \cdot (\vec{g}_2 \times \vec{g}_3)}{v} = 1 \tag{1.1.6a}$$

Multiplying both sides of Eq. (1.1.4) into $\vec{g}_2$ considering Eq. (1.1.2) provides:

$$\vec{g}_2 \cdot \vec{g}^2 = \frac{\vec{g}_2 \cdot (\vec{g}_1 \times \vec{g}_3)}{v} = 1 \tag{1.1.6b}$$

Multiplying both sides of Eq. (1.1.5) into $\vec{g}_3$ considering Eq. (1.1.2) provides:

$$\vec{g}_3 \cdot \vec{g}^3 = \frac{\vec{g}_3 \cdot (\vec{g}_1 \times \vec{g}_2)}{v} = 1 \tag{1.1.6c}$$

It seems that contra-variant base vector system is the answer to the question, because if we multiply both sides of equation (1.1.1) into $\vec{g}^1$, $\vec{g}^2$ and $\vec{g}^3$, respectively, considering the equations in Eq. (1.1.6) then we have:

$$x^1 = (\vec{r} \cdot \vec{g}^1) \quad x^2 = (\vec{r} \cdot \vec{g}^2) \quad x^3 = (\vec{r} \cdot \vec{g}^3) \tag{1.1.7}$$

We can conclude that in a non-orthogonal coordinates we face a dual system of base vector the first one is covariant base vectors $\vec{g}_1$, $\vec{g}_2$ and $\vec{g}_3$, and 2nd one is the contravariant base vectors $\vec{g}^1$, $\vec{g}^2$ and $\vec{g}^3$.

Any vector $\vec{r}$ can be represented in covariant base vectors with contravariant components such that:

$$\vec{r} = x^1\,\vec{g}_1 + x^2\,\vec{g}_2 + x^3\,\vec{g}_3 \tag{1.1.1}$$

Alternatively any vector $\vec{r}$ can be represented in contravariant base vectors with covariant components such that:

$$\vec{r} = x_1\,\vec{g}^1 + x_2\,\vec{g}^2 + x_3\,\vec{g}^3 \tag{1.1.8}$$

Obviously the covariant components are:

$$x_1 = (\vec{r} \cdot \vec{g}_1) \quad x_2 = (\vec{r} \cdot \vec{g}_2) \quad x_3 = (\vec{r} \cdot \vec{g}_3) \tag{1.1.9}$$

Orthogonal base vector systems are special case of the general non-orthogonal system in which the covariant and contra-variant base vectors are coincide with each other. For example, in a cylindrical orthogonal base vectors we have:

$$\vec{g}_\theta = \vec{g}^\theta = \vec{e}_\theta \qquad |\vec{e}_\theta| = 1 \qquad \vec{g}_r = \vec{g}^r = \vec{e}_r \qquad |\vec{e}_r| = 1$$

$$\vec{g}_z = \vec{g}^z = \vec{e}_z \qquad |\vec{e}_z| = 1$$

1.2 METRIC TENSORS AND THE APPLICATIONS

In Section 1.1. we mentioned that for non-orthogonal coordinates a dual system consist of covariant and contra-variant base vector system is required. Any position vector $\vec{r}$ can be expressed by covariant base vectors like Eq. (1.1.1) or contra-variant base vectors like Eq. (1.1.8).

If we substitute Eq. (1.1.9) into Eq. (1.1.8), the by using abbreviated summation index j, we have:

$$\vec{r} = \left(\vec{r} \cdot \vec{g}_j\right)\vec{g}^j \tag{1.2.1}$$

Similarly by substituting Eq. (1.1.7) into Eq. (1.1.1), and using summation index j, we have:

$$\vec{r} = \left(\vec{r} \cdot \vec{g}^j\right)\vec{g}_j \tag{1.2.2}$$

From now on for sake of simplicity we remove the vector sign from the base vectors saying that $\vec{g}^i = g^i$ and also $\vec{g}_i = g_i$. Now if we set $\vec{r} = g_i$ in Eq. (1.2.1) and $\vec{r} = g^j$ in Eq. (1.2.2) and remove vector signs we have:

$$g_i = \left(g_i \cdot g_j\right)g^j \qquad g^i = \left(g^i \cdot g^j\right)g_j \tag{1.2.3}$$

The pair of equations in Eq. (1.2.3) provides a relationship between covariant and contra-variant vector base system and the abbreviated form is:

$$g_i = g_{ij}\, g^j \tag{1.2.4}$$

$$g^i = g^{ij}\, g_j \tag{1.2.5}$$

$$g_{ij} = \left(g_i \cdot g_j\right) \qquad g^{ij} = \left(g^i \cdot g^j\right) \tag{1.2.6}$$

The g_{ij} and g^{ij} defined by Eq. (1.2.6), are called metric tensors which obviously relates the covariant to contra-variant base vectors and vice versa. This tensor has an important role in calculations of the strain and stress in finite deformations. For an orthogonal base vector system it is obvious that $i, j = x, y, z \quad g_{ij} = g^{ij} = \delta_{ij}$ and the δ_{ij} is called Kronecker delta defined by:

$$\delta_{ij} = 1 \quad i = j \qquad \delta_{ij} = 0 \quad i \neq j$$

If we multiply (scalar product) both sides of Eq. (1.2.4) into the base vector g^k then we have:

$$g_i \cdot g^k = g_{ij}\, g^j \cdot g^k$$

Rewriting the expressions in Eq. (1.1.6) in compact form is like this:

$$g_i \cdot g^k = \delta_i^k \tag{1.1.6}$$

Kronecker delta δ_i^k herein is:

$$\delta_i^k = 1 \quad k = i \quad \delta_i^k = 0 \quad k \neq i$$

Since $g^j \cdot g^k = g^{jk}$ (see Eq. (1.2.6)) the by substituting Eqs. (1.1.6) and Eq. (1.2.6) into $g_i \cdot g^k = g_{ij}\, g^j \cdot g^k$ we have:

$$g_{ij} \cdot g^{jk} = \delta_i^k \tag{1.2.7}$$

We can use metric tensor to find out relationship between covariant and contra-variant components of a vector since we can write:

$$\vec{r} = x^i\, g_i = x_i\, g^i \tag{1.2.8}$$

If we substitute Eqs. (1.2.4) and (1.2.5) into Eq. (1.2.8) then we have, $\vec{r} = x^i\, g_{ij}\, g^j = x_j\, g^j$ and also $\vec{r} = x^j\, g_j = x_i\, g^{ij}\, g_j$ and this results two important formulas:

$$x^j = x_i\, g^{ij} \quad x_j = x^i\, g_{ij} \tag{1.2.9}$$

The Eq. (1.2.9) is for components of a vector, similarly can be valid for components of a tensor $\boldsymbol{A}$ as follows:

$$\boldsymbol{A} = A^{kl} g_k\, g_l = A^{11} g_1\, g_1 + A^{12} g_1\, g_2 + A^{13} g_1\, g_3 + A^{21} g_2\, g_1 + A^{22} g_2\, g_2 + A^{23} g_2\, g_3 + A^{31} g_3\, g_1 + A^{32} g_3\, g_2 + A^{33} g_3\, g_3 \tag{1.2.10}$$

In Eq. (1.2.10) the product $g_k\, g_l$ (without dot between) is called outer product and its result is a 3×3 matrix as follows:

$$g_k\, g_l = g_k \otimes g_l = \begin{bmatrix} 0 & 0 & 0 \\ 0 & 0 & g_{kl} \\ 0 & 0 & 0 \end{bmatrix}$$

Equation (1.2.10) is covariant-based expression of a tensor **A**, similarly we can write the contra-variant expression as follows:

$$\boldsymbol{A} = A_{ij}g^i\, g^j = A_{11}g^1\, g^1 + A_{12}g^1\, g^2 + A_{13}g^1\, g^3 + A_{21}g^2\, g^1 + A_{22}g^2\, g^2$$
$$+ A_{23}g^2\, g^3 + A_{31}g^3\, g^1 + A_{32}g^3\, g^2 + A_{33}g^3\, g^3 \qquad (1.2.11)$$

In Eq. (1.2.11) the product $g^i\, g^j$ (without dot between) is called outer product and its result is a 3×3 matrix as follows:

$$g^1\, g^2 = g^1 \otimes g^2 = \begin{bmatrix} 0 & g^{12} & 0 \\ 0 & 0 & 0 \\ 0 & 0 & 0 \end{bmatrix} \qquad g^3\, g^3 = g^3 \otimes g^3 = \begin{bmatrix} 0 & 0 & 0 \\ 0 & 0 & 0 \\ 0 & 0 & g^{33} \end{bmatrix}$$

According to Eqs. (1.2.10) and (1.2.11), any tensor can be expressed either in terms of covariant or contra-variant base vector and the abbreviated form is:

$$\boldsymbol{A} = A^{kl}g_k\, g_l = A_{ij}g^i\, g^j \qquad i,j,k,l = 1,2,3 \qquad (1.2.12)$$

From Eq. (1.2.4) we have $g_k = g_{ki}\, g^i$ and also $g_l = g_{ji}\, g^j$ and if we substitute these in Eq. (1.2.12) we have:

$$A_{ij} = A^{kl} g_{ki}\, g_{lj} \qquad (1.2.13)$$

From Eq. (1.2.5) we have $g^i = g^{ik}\, g_k$ and also $g^j = g^{jl}\, g_l$ and if we substitute these in Eq. (1.2.12) we have:

$$A^{kl} = g^{ki}\, g^{lj} A_{ij} \qquad (1.2.14)$$

Therefore, the main role of the metric tensor is relating covariant and contra-variant components of a tensor (see Eqs. (1.2.13) and (1.2.14)) and vectors (see Eq. (1.2.9)). If we have two vectors $\vec{u}$ and $\vec{v}$, then their scalar product is:

$$\vec{u}\cdot\vec{v} = \left(u_i\, g^i\right)\cdot\left(v_j\, g^j\right) = u_i v_j g^i \cdot g^j = u_i v_j g^{ij} = u^j v_j \qquad (1.2.15)$$

In Eq. (1.2.15) u^j is contra-variant component of $\vec{u}$ and v_j is covariant component of $\vec{v}$, it is worth reminding that they are not physical length of the vector, since in non-orthogonal coordinate systems the base vectors do not have unit length and components cannot represent the length.

For example a covariant base vector g_i is not a unit vector, but a unit vector e_i across a base vector g_i can be easily calculated from the following formula:

$$e_i = \frac{g_i}{|g_i|} = \frac{g_i}{\sqrt{g_{ii}}} \tag{1.2.16}$$

The above formula enables a vector to be expressed in terms of e_i with the components u^{*i} which is physical length of the vector, the formula is:

$$\vec{u} = u^i g_i = u^{*i} e_i \tag{1.2.17}$$

If we substitute Eqs. (1.2.16) into (1.2.17) then the physical length can be found, that is,

$$u^{*i} = \sqrt{g_{ii}}\, u^i \tag{1.2.18}$$

Similarly for the contra-variant system a formula can be developed in the same way which is:

$$u_i^* = \sqrt{g^{ii}}\, u_i \tag{1.2.19a}$$

The expanded versions of g_1, g_2 and g_3 are:

$$g_1 = \frac{\partial \vec{r}}{\partial \alpha_1} = \frac{\partial x_1}{\partial \alpha_1}\vec{i}_1 + \frac{\partial x_2}{\partial \alpha_1}\vec{i}_2 + \frac{\partial x_3}{\partial \alpha_1}\vec{i}_3 \quad g_2 = \frac{\partial \vec{r}}{\partial \alpha_2} = \frac{\partial x_1}{\partial \alpha_2}\vec{i}_1 + \frac{\partial x_2}{\partial \alpha_2}\vec{i}_2 + \frac{\partial x_3}{\partial \alpha_2}\vec{i}_3$$

$$g_3 = \frac{\partial \vec{r}}{\partial \alpha_3} = \frac{\partial x_1}{\partial \alpha_3}\vec{i}_1 + \frac{\partial x_2}{\partial \alpha_3}\vec{i}_2 + \frac{\partial x_3}{\partial \alpha_3}\vec{i}_3$$

In the last section, we mentioned that volume of this parallelepiped formed by covariant base vector g_1, g_2 and g_3 is:

$$v = g_1 \cdot (g_2 \times g_3) = g_2 \cdot (g_1 \times g_3) = g_3 \cdot (g_1 \times g_2)$$

The triple vector products above can be expressed by a determinant which is:

$$g_3 \cdot (g_1 \times g_2) = \begin{vmatrix} \dfrac{\partial x_1}{\partial \alpha_1} & \dfrac{\partial x_2}{\partial \alpha_1} & \dfrac{\partial x_3}{\partial \alpha_1} \\ \dfrac{\partial x_1}{\partial \alpha_2} & \dfrac{\partial x_2}{\partial \alpha_2} & \dfrac{\partial x_3}{\partial \alpha_2} \\ \dfrac{\partial x_1}{\partial \alpha_3} & \dfrac{\partial x_2}{\partial \alpha_3} & \dfrac{\partial x_3}{\partial \alpha_3} \end{vmatrix} \tag{1.2.19b}$$

To expand elements of the metric tensor g_{ij} say for example g_{12} we can write:

$$g_{12} = g_1 \cdot g_2 = \left(\frac{\partial x_1}{\partial \alpha_1} \vec{i}_1 + \frac{\partial x_2}{\partial \alpha_1} \vec{i}_2 + \frac{\partial x_3}{\partial \alpha_1} \vec{i}_3 \right) \cdot \left(\frac{\partial x_1}{\partial \alpha_2} \vec{i}_1 + \frac{\partial x_2}{\partial \alpha_2} \vec{i}_2 + \frac{\partial x_3}{\partial \alpha_2} \vec{i}_3 \right)$$

Since the scalar products $\vec{i}_1 \cdot \vec{i}_2 = \vec{i}_2 \cdot \vec{i}_3 = \vec{i}_3 \cdot \vec{i}_1 = 0$ and $\vec{i}_1 \cdot \vec{i}_1 = \vec{i}_2 \cdot \vec{i}_2 = \vec{i}_3 \cdot \vec{i}_3 = 1$, the above expression can be simplified to:

$$g_{12} = \frac{\partial x_1}{\partial \alpha_1} \frac{\partial x_1}{\partial \alpha_2} + \frac{\partial x_2}{\partial \alpha_1} \frac{\partial x_2}{\partial \alpha_2} + \frac{\partial x_3}{\partial \alpha_1} \frac{\partial x_3}{\partial \alpha_2}$$

Abbreviation of the above expression by using summation index yields to:

$$g_{ij} = \frac{\partial x_m}{\partial \alpha_i} \frac{\partial x_m}{\partial \alpha_j} \tag{1.2.20}$$

By looking at the above expression it can be realized easily that the matrix $[\mathbf{g}]$ containing the elements g_{ij} is a result of product of a matrix and its transpose as follows:

$$[\mathbf{g}] = \begin{bmatrix} \dfrac{\partial x_1}{\partial \alpha_1} & \dfrac{\partial x_2}{\partial \alpha_1} & \dfrac{\partial x_3}{\partial \alpha_1} \\ \dfrac{\partial x_1}{\partial \alpha_2} & \dfrac{\partial x_2}{\partial \alpha_2} & \dfrac{\partial x_3}{\partial \alpha_2} \\ \dfrac{\partial x_1}{\partial \alpha_3} & \dfrac{\partial x_2}{\partial \alpha_3} & \dfrac{\partial x_3}{\partial \alpha_3} \end{bmatrix} \begin{bmatrix} \dfrac{\partial x_1}{\partial \alpha_1} & \dfrac{\partial x_1}{\partial \alpha_2} & \dfrac{\partial x_1}{\partial \alpha_3} \\ \dfrac{\partial x_2}{\partial \alpha_1} & \dfrac{\partial x_2}{\partial \alpha_2} & \dfrac{\partial x_2}{\partial \alpha_3} \\ \dfrac{\partial x_3}{\partial \alpha_1} & \dfrac{\partial x_3}{\partial \alpha_2} & \dfrac{\partial x_3}{\partial \alpha_3} \end{bmatrix} \tag{1.2.21}$$

In linear algebra we can prove that for two square matrices $[\boldsymbol{B}]$ and $[\boldsymbol{C}]$, when their product is defined by $[\boldsymbol{A}]=[\boldsymbol{B}][\boldsymbol{C}]$, then their determinants also related by $|\boldsymbol{A}|=|\boldsymbol{B}||\boldsymbol{C}|$. If we apply this theorem to the matrix product in Eq. (1.2.21) by considering that the determinant in Eq. (1.2.19b) is the same of determinant of its transpose then by considering $|\boldsymbol{g}|=\det(g_{ij})=g$, we have:

$$|\boldsymbol{g}|=\det(g_{ij})=(g_3\cdot(g_1\times g_2))(g_3\cdot(g_1\times g_2))=g$$

Therefore we have the following formula:

$$g_3\cdot(g_1\times g_2)=\sqrt{g} \quad (1.2.22)$$

1.3 ELEMENTARY OPERATIONS WITH METRIC TENSORS

In this chapter, we discuss about differentials of the length, area and volume also we discuss about covariant differentiation. These concepts are necessary for using tensors in nonlinear solid mechanics. From previous section, the concept of metric tensors and Eq. (1.2.22) will be used throughout the section. First we study the length differential according to Fig. 1.4, we have:

$$ds \cong |d\vec{r}|$$

$$ds^2=d\vec{r}\cdot d\vec{r}$$

From Eq. (1.1.1) we have:

$d\vec{r}=g_1\,dx^1+g_2\,dx^2+g_3\,dx^3$ which yields to:

$$ds^2=(g_1\,dx^1+g_2\,dx^2+g_3\,dx^3)\cdot(g_1\,dx^1+g_2\,dx^2+g_3\,dx^3)$$

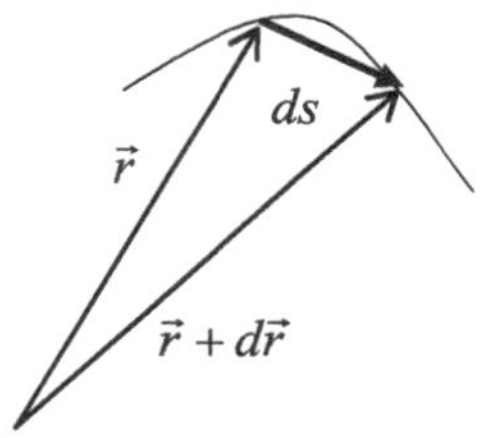

FIGURE 1.4 Length differential.

The above expression consists of nine terms and it can be abbreviated by using Eq. (1.2.6) into the following form:

$$ds^2 = g_{ij}\, dx^i\, dx^j \qquad i.j = 1,2,3 \tag{1.3.1}$$

From Eq. (1.3.1) we see that g_{ij} is a prelude for calculating the length and the term "metric tensor" for it is used due to this fact. The elements of area can also be represented by a vector products in Fig. 1.5, the surfaces α_1, α_2 and α_3 are shown and any differential area vector $d\vec{a}_3$ and be by following vector product:

$$d\vec{a}_3 = g_1\, dx^1 \times g_2\, dx^2$$

This is a definition and to be consistent with orthogonal case it can be written into this form:

$$d\vec{a}_3 = (g_1 \times g_2)\, dx^1\, dx^2 \tag{1.3.2}$$

In orthogonal system $(g_1 \times g_2)$ is a unit vector, therefore, Eq. (1.3.2) is a general definition.

From Eq. (1.1.5) we know that $(g_1 \times g_2)$ lies in g^3 direction, therefore it is necessary to find the scalar product of $e^3 \cdot (g_1 \times g_2)$ to find out the $|d\vec{a}_3|$ considering that $e^3 = \dfrac{g^3}{\sqrt{g^{33}}}$ (see Eq. (1.2.16))

$$|d\vec{a}_3| = d\,a_3 = d\vec{a}_3 \cdot \frac{g^3}{\sqrt{g^{33}}}$$

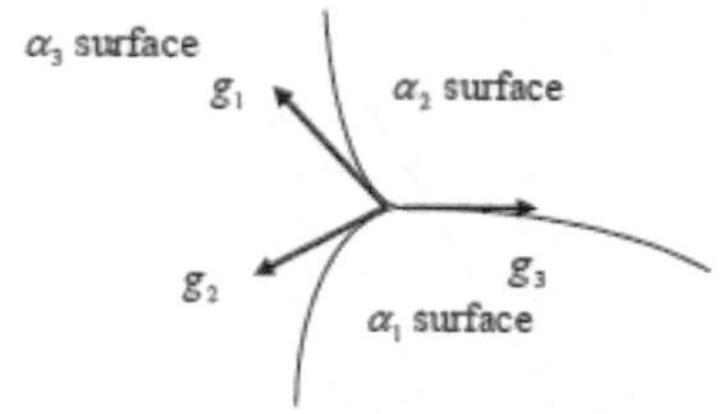

FIGURE 1.5 Surfaces α_1, α_2 and α_3.

The numerical value of $d\vec{a}_3$ is $d\,a_3$ and if we substitute Eq. (1.3.2) into above expression we have:

$$d\,a_3 = (g_1 \times g_2)\cdot \frac{g^3}{\sqrt{g^{33}}}\,dx^1\,dx^2 \qquad (1.3.3)$$

From Eq. (1.2.5) we can conclude that:

$$g^3 = g^{3j}\,g_j \qquad (1.3.4)$$

If we substitute Eqs. (1.3.4) into (1.3.3) and consider that when $j = 1, 2$ we have:

$$g_1 \cdot (g_1 \times g_2) = g_2 \cdot (g_1 \times g_2) = 0$$

Because both g_1 and g_2 are perpendicular to $(g_1 \times g_2)$ and only for $j = 3$ we have nonzero terms. Therefore, Eq. (1.3.3) will change to:

$$d\,a_3 = (g_1 \times g_2)\cdot g^3 \sqrt{g^{33}}\,dx^1\,dx^2$$

Now we need to substitute Eq. (1.2.22) into the above equation, we find out that:

$$d\,a_3 = \sqrt{g\,g^{33}}\,dx^1\,dx^2 \qquad (1.3.5)$$

Similarly we can find the expressions for both $d\,a_1$ and $d\,a_2$

$$d\,a_1 = \sqrt{g\,g^{11}}\,dx^3\,dx^2 \qquad (1.3.6)$$

$$d\,a_2 = \sqrt{g\,g^{22}}\,dx^1\,dx^3 \qquad (1.3.7)$$

We can also calculate volume differential by considering the parallelepiped that expressed in Eq. (1.1.2) as follows:

$$v = \vec{g}_1 \cdot (\vec{g}_2 \times \vec{g}_3) = \vec{g}_2 \cdot (\vec{g}_1 \times \vec{g}_3) = \vec{g}_3 \cdot (\vec{g}_1 \times \vec{g}_2) \qquad (1.1.2)$$

Volume differential is still a parallelepiped that can be described by:

$$dv = d\vec{a}_3 \cdot d\vec{g}_3$$

This can be expended by substituting Eq. (1.3.2) into it, that is,

$$dv = (g_1 \times g_2)dx^1\, dx^2 \cdot g_3\, dx^3 = (g_1 \times g_2)\cdot g_3\, dx^1\, dx^2 dx^3$$

Then by using Eq. (1.2.22) the above expression changes to:

$$dv = \sqrt{g}\ dx^1\, dx^2 dx^3 \tag{1.3.8}$$

Which is a general formula, for volume differential and for an orthogonal coordinate $\sqrt{g} = 1$.

Now we talk about natural differentiation of a vector. It is designated by $\vec{u}_{,i}$ which is $\dfrac{\partial \vec{u}}{\partial \alpha_i}$, the vector $\vec{u}$ is, $\vec{u} = u^1 g_1 + u^2 g_2 + u^3 g_3$, therefore the differentiation will be:

$$\vec{u}_{,i} = \frac{\partial \vec{u}}{\partial \alpha_i} = \frac{\partial u^1}{\partial \alpha_i} g_1 + u^1 \frac{\partial g_1}{\partial \alpha_i} + \frac{\partial u^2}{\partial \alpha_i} g_2 + u^2 \frac{\partial g_2}{\partial \alpha_i} + \frac{\partial u^3}{\partial \alpha_i} g_3 + u^3 \frac{\partial g_3}{\partial \alpha_i} \tag{1.3.9}$$

Equation (1.3.9) consists of six terms. Now we define Christoffel symbols as follows:

$$\frac{\partial g_1}{\partial \alpha_i} = \Gamma^r_{1i}\, g_r \qquad \frac{\partial g_2}{\partial \alpha_i} = \Gamma^r_{2i}\, g_r \qquad \frac{\partial g_3}{\partial \alpha_i} = \Gamma^r_{3i}\, g_r \tag{1.3.10}$$

Each expression in Eq. (1.3.10) will be expanded into 3 terms, therefore if Eq. (1.3.10) substituted in Eq. (1.3.9) and then expanded 12 terms will in the expansion like this:

$$\vec{u}_{,i} = \left(\frac{\partial u^1}{\partial \alpha_i} + \Gamma^1_{1i}\, u^1 + \Gamma^1_{2i}\, u^2 + \Gamma^1_{3i}\, u^3 \right) g_1 + \left(\frac{\partial u^2}{\partial \alpha_i} + \Gamma^2_{1i}\, u^1 + \Gamma^2_{2i}\, u^2 + \Gamma^2_{3i}\, u^3 \right) g_2$$
$$+ \left(\frac{\partial u^3}{\partial \alpha_i} + \Gamma^3_{1i}\, u^1 + \Gamma^3_{2i}\, u^2 + \Gamma^3_{3i}\, u^3 \right) g_1 \tag{1.3.11}$$

The natural differentiation displayed in Eq. (1.3.11) consists of 12 terms as above. However, it should be simplified and the brackets that are the components of g_1, g_2 and g_3 can be designated by $u^1\big|_i$, $u^2\big|_i$ and $u^3\big|_i$, this enables simplified form of Eq. (1.3.11) to be written in three terms as indicated by:

$$\vec{u}_{,i} = u^1\big|_i\, g_1 + u^2\big|_i\, g_2 + u^3\big|_i\, g_3 \tag{1.3.12}$$

Comparing Eqs. (1.3.12) and (1.3.11) introduces terms $u^1\big|_i$, $u^2\big|_i$ and $u^3\big|_i$ which are known as covariant differentiation of the components of a vector. These three terms are components of g_1, g_2 and g_3 respectively and make the vector differentiation simplified. According to Eqs. (1.3.11) and (1.3.12) the covariant differentiation of the components are:

$$u^1\big|_i = \underbrace{\frac{\partial u^1}{\partial \alpha_i} + \Gamma^1_{ki}\, u^k}_{4\ terms} \qquad u^2\big|_i = \underbrace{\frac{\partial u^2}{\partial \alpha_i} + \Gamma^2_{ki}\, u^k}_{4\ terms} \qquad u^3\big|_i = \underbrace{\frac{\partial u^3}{\partial \alpha_i} + \Gamma^3_{ki}\, u^k}_{4\ terms} \tag{1.3.13}$$

Equations (1.3.13) can be shown in an abbreviated form in one of the two forms and both are acceptable for definition of the covariant differentiation they are:

$$u^j\big|_i = \frac{\partial u^j}{\partial \alpha_i} + \Gamma^j_{ki}\, u^k \qquad u^i\big|_j = \frac{\partial u^i}{\partial \alpha_j} + \Gamma^i_{kj}\, u^k \tag{1.3.14}$$

Now it is time to explain what is Γ^i_{kj} $\; i,j,k = 1,2,3$. They are called Christoffel symbols (27 in total) and are necessary for determination of the covariant differentiation of the components of a vector.

In order to find an explicit formula for them we form the scalar product of g^1 and 1st in Eq. (1.3.10), that is, $g^1 \cdot \dfrac{\partial g_1}{\partial \alpha_i} = g^1 \cdot \Gamma^r_{1i}\, g_r$ and considering $g^1 \cdot g_1 = 1$ also $g^1 \cdot g_2 = g^1 \cdot g_3 = 0$, then we have:

$$\Gamma^1_{1i} = g^1 \cdot \frac{\partial g_1}{\partial \alpha_i} \qquad i = 1,2,3$$

The above shows three symbols. By forming the scalar product of g^2 and g^3 to 1st in Eq. (1.3.10) and considering $g^2 \cdot g_2 = g^3 \cdot g_3 = 1$ also $g^2 \cdot g_r = 0 \quad r = 1,3$ and $g^3 \cdot g_r = 0 \quad r = 1,2$, then we have further six symbols like this:

$$\Gamma_{1i}^2 = g^2 \cdot \frac{\partial g_1}{\partial \alpha_i} \quad i = 1,2,3 \qquad \Gamma_{1i}^3 = g^3 \cdot \frac{\partial g_1}{\partial \alpha_i} \quad i = 1,2,3$$

Further nine symbols can be found by forming the scalar product of g^1, g^2 and g^3 to 2nd in Eq. (1.3.10) and considering $g^1 \cdot g_1 = g^2 \cdot g_2 = g^3 \cdot g_3 = 1$ also $g^1 \cdot g_r = 0 \quad r = 2,3$ $g^2 \cdot g_r = 0 \quad r = 1,3$ and $g^3 \cdot g_r = 0 \quad r = 1,2$, they are:

$$\Gamma_{2i}^1 = g^1 \cdot \frac{\partial g_2}{\partial \alpha_i} \quad i = 1,2,3 \qquad \Gamma_{2i}^2 = g^2 \cdot \frac{\partial g_2}{\partial \alpha_i} \quad i = 1,2,3$$

$$\Gamma_{2i}^3 = g^3 \cdot \frac{\partial g_2}{\partial \alpha_i} \quad i = 1,2,3$$

The final nine symbols can be found by forming the scalar product of g^1, g^2 and g^3 to 3rd in Eq. (1.3.10) and considering $g^1 \cdot g_1 = g^2 \cdot g_2 = g^3 \cdot g_3 = 1$ also $g^1 \cdot g_r = 0 \quad r = 2,3$ $g^2 \cdot g_r = 0 \quad r = 1,3$ and $g^3 \cdot g_r = 0 \quad r = 1,2$, they are:

$$\Gamma_{3i}^1 = g^1 \cdot \frac{\partial g_3}{\partial \alpha_i} \quad i = 1,2,3 \qquad \Gamma_{3i}^2 = g^2 \cdot \frac{\partial g_3}{\partial \alpha_i} \quad i = 1,2,3$$

$$\Gamma_{3i}^3 = g^3 \cdot \frac{\partial g_3}{\partial \alpha_i} \quad i = 1,2,3$$

All the above 27 symbols can be summarized in one formula like this:

$$\Gamma_{ji}^k = g^k \cdot \frac{\partial g_j}{\partial \alpha_i} \quad i,j,k = 1,2,3 \tag{1.3.15}$$

They are Christoffel symbols and should be used to find the covariant differentiation of the components of a vector (rewriting Eq. (1.3.14)):

$$u^j\big|_i = \frac{\partial u^j}{\partial \alpha_i} + \Gamma_{ki}^j u^k$$

The covariant differentiation of the components, are used to express the natural differentiation of vector (rewrite Eq. (1.3.12)), that is,

$$\vec{u}_{,i} = u^j\big|_i \, g_j \qquad i,j = 1,2,3 \tag{1.3.16}$$

KEYWORDS

- **Christoffel symbols**
- **covariant differentiation**
- **metric tensors**
- **non-orthogonal coordinate system.**
- **orthogonal coordinate system**

CHAPTER 2

GENERAL THEORY FOR DEFORMATION AND STRAIN IN SOLIDS

CONTENTS

2.1 STRAIN DEFINITION IN LARGE DEFORMATIONS

A well-known criteria that we remember from engineering and represents deformation, is named strain. Defining engineering strain is very simple, if we imagine a rod with an initial length l_0 (see Fig. 2.1), when subjected to tension its length increases to l_n, then the engineering strain is defined by:

$$\varepsilon_E = \frac{l_n - l_0}{l_0} \tag{2.1.1}$$

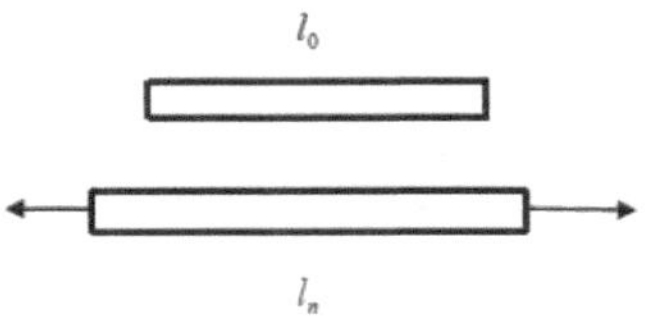

FIGURE 2.1 Rod under tension.

If we multiply the numerator and denominator of Eq. (2.1.1) into $l_0 + l_n$ then we have:

$$\varepsilon_E = \frac{(l_n - l_0)(l_n + l_0)}{(l_n + l_0)l_0} = \frac{l_n^2 - l_0^2}{l_0^2\left(1 + \frac{l_n}{l_0}\right)} = \frac{l_n^2 - l_0^2}{l_0^2\left(2 + \frac{l_n - l_0}{l_0}\right)} = \frac{l_n^2 - l_0^2}{l_0^2(2 + \varepsilon_E)}$$

In the above expression if $\varepsilon_E << 2$, then we can ignore it and define a new strain known as "Green strain" which will be:

$$\varepsilon_G = \frac{l_n^2 - l_0^2}{2l_0^2} \tag{2.1.2}$$

Comparing Eqs. (2.1.1) and (2.1.2) shows that $\varepsilon_E = \frac{\varepsilon_G}{\left(1 + \varepsilon_E/2\right)}$ or alternatively:

$$\varepsilon_G = \varepsilon_E\left(1 + \varepsilon_E/2\right)$$

It is obvious that if $\varepsilon_E << 1$ (small deformation), in Eq. (3) we can ignore the term $\varepsilon_E^2/2$ and then we can say that Green strain ε_G and engineering strain ε_E are close $\varepsilon_G \cong \varepsilon_E$. Meanwhile we can rearrange Eq. (2.1.2) into the following form:

$$l_n^2 - l_0^2 = 2l_0^2\,\varepsilon_G \tag{2.1.3}$$

When the deformation is large and is not a uni-axial tension type, we can also define the Green strain tensor by relaying on a simple definition in Eq. (2.1.3). According to Fig. 2.2, vector $\vec{r}$ shows position of a point in un-deformed position and the metric tensor can be shown by $\boldsymbol{g}$, the differential length element ds can be found from Eq. (1.3.1) written by:

$$ds^2 = g_{ij}\,dx^i\,dx^j \qquad i.j = 1,2,3 \tag{2.1.4}$$

In the deformed body vector $\vec{R}$ is position of the same point and the metric tensor will change to $\boldsymbol{G}$ because the deformation may be too big that

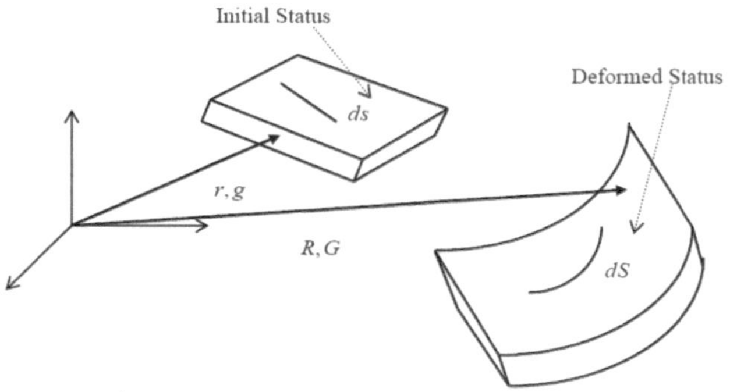

FIGURE 2.2 Deformed and un-deformed geometry.

causes change of the natural coordinates. Similarly the differential length element dS can be found from Eq. (1.3.1) written by:

$$dS^2 = G_{ij}\, dx^i\, dx^j \qquad i.j = 1,2,3 \tag{2.1.5}$$

General form of Eq. (2.1.3) gives Green strain tensor like this:

$$dS^2 - ds^2 = 2\, \varepsilon_{ij} dx^i dx^j \tag{2.1.6}$$

By substituting Eqs. (2.1.4) and (2.1.5) into Eq. (2.1.6) we have:

$$G_{ij}\, dx^i\, dx^j - g_{ij}\, dx^i\, dx^j = 2\, \varepsilon_{ij} dx^i dx^j$$

Simplify the above equation gives the Green strain tensor which is:

$$\varepsilon_{ij} = \frac{1}{2}\left(G_{ij} - g_{ij}\right) \tag{2.1.7}$$

Equation (2.1.7) shows that Green strain tensor has covariant components; therefore, it should be expressed by contra-variant base vectors as indicated in Eq. (1.2.11) ε_{ij} can be expressed by the un-deformed contra-variant base vectors like this:

$$\begin{aligned} \varepsilon = \varepsilon_{ij} g^i\, g^j &= \varepsilon_{11} g^1\, g^1 + \varepsilon_{12} g^1\, g^2 + \varepsilon_{13} g^1\, g^3 + \varepsilon_{21} g^2\, g^1 + \varepsilon_{22} g^2\, g^2 + \\ &+ \varepsilon_{23} g^2\, g^3 + \varepsilon_{31} g^3\, g^1 + \varepsilon_{32} g^3\, g^2 + \varepsilon_{33} g^3\, g^3 \end{aligned} \tag{2.1.8}$$

According to Fig. 2.2, the displacement vector $\vec{u}$ is:

$$\vec{R} = \vec{r} + \vec{u}$$

If we differentiate versus the coordinate α_i, then we have $\frac{\partial \vec{R}}{\partial \alpha_i} = \frac{\partial \vec{r}}{\partial \alpha_i} + \frac{\partial \vec{u}}{\partial \alpha_i}$ since $\frac{\partial \vec{R}}{\partial \alpha_i} = G_i$, $\frac{\partial \vec{r}}{\partial \alpha_i} = g_i$ and $\frac{\partial \vec{u}}{\partial \alpha_i} = \vec{u},_i$ then we can write:

$$G_i = g_i + \vec{u},_i \qquad G_j = g_j + \vec{u},_j \qquad G_{ij} = G_i \cdot G_j = (g_i + \vec{u},_i) \cdot (g_j + \vec{u},_j) \tag{2.1.9}$$

By substituting Eq. (2.1.9) and also $g_{ij} = g_i \cdot g_j$ into Eq. (2.1.7) the Green strain tensor can be written in terms of displacement vector as follows:

$$\varepsilon_{ij} = \frac{1}{2}\left(g_i \cdot \vec{u},_j + g_j \cdot \vec{u},_i + \vec{u},_i \cdot \vec{u},_j\right) \tag{2.1.10}$$

It is also possible that express the Green strain tensor in terms of the deformed covariant base vectors, that is, substitute $g_{ij} = g_i \cdot g_j = (G_i - \vec{u},_i) \cdot (G_j - \vec{u},_j)$ and also $G_{ij} = G_i \cdot G_j$ into Eq. (2.1.7) to have the following expression:

$$\varepsilon_{ij} = \frac{1}{2}\left(G_i \cdot \vec{u},_j + G_j \cdot \vec{u},_i - \vec{u},_i \cdot \vec{u},_j\right) \tag{2.1.11}$$

It is also possible to use covariant differentiation of the displacement components from Eq. (1.3.16) to change the strain formulas for strain tensor, that is,

$$\vec{u} = u_m g^m \qquad \vec{u},_i = u_m|_i \, g^m \tag{2.1.12a}$$

$$\vec{u} = U_m G^m \qquad \vec{u},_i = U_m|_i \, G^m \tag{2.1.12b}$$

It should be remembered that Eq. (1.3.16) is written in terms of covariant base vectors while Eqs. (2.1.12a) and (2.1.12b) are written in terms of contra-variant base vectors. This is because Green strain is a covariant tensor and needs an expression with covariant components.

Substituting Eq. (2.1.12a) into Eq. (2.1.10) a term like $g_i \cdot u_m|_j g^m$ appears and should be simplified.

Exercise 2.1.1: *Simplify* $g_i \cdot u_m|_j g^m$

Solution:
$g_i \cdot u_m|_j g^m = g_i \cdot u_1|_j g^1 + g_i \cdot u_1|_j g^1 + g_i \cdot u_3|_j g^3$, when we consider that $g^1 \cdot g_1 = g^2 \cdot g_2 = g^3 \cdot g_3 = 1$ also $g^1 \cdot g_r = 0$ $r = 2,3$ $g^2 \cdot g_r = 0$ $r = 1,3$ and $g^3 \cdot g_r = 0$ $r = 1,2$, then we have:

$$g_i \cdot u_1|_j g^1 + g_i \cdot u_1|_j g^1 + g_i \cdot u_3|_j g^3 = u_1|_j \quad i = 1$$
$$g_i \cdot u_1|_j g^1 + g_i \cdot u_1|_j g^1 + g_i \cdot u_3|_j g^3 = u_2|_j \quad i = 2$$
$$g_i \cdot u_1|_j g^1 + g_i \cdot u_1|_j g^1 + g_i \cdot u_3|_j g^3 = u_3|_j \quad i = 3$$

Therefore we have:

$$g_i \cdot u_m|_j g^m = u_i|_j \qquad g_j \cdot u_m|_i g^m = u_j|_i \tag{2.1.12c}$$

Exercise 2.1.2: *Simplify* $\vec{u},_i \cdot \vec{u},_j$

Solution: Consider $\vec{u},_i = u^1|_i g_1 + u^2|_i g_2 + u^3|_i g_3$ but it can also be expressed in contra-variant base vector, that is, $\vec{u},_j = u_1|_j g^1 + u_2|_j g^2 + u_3|_j g^3$, therefore we have:

$\vec{u},_i \cdot \vec{u},_j = \left(u^1|_i g_1 + u^2|_i g_2 + u^3|_i g_3\right) \cdot \left(\vec{u},_j = u_1|_j g^1 + u_2|_j g^2 + u_3|_j g^3\right)$, this product have nine terms in it but it can be reduced to three terms only because $g^1 \cdot g_1 = g^2 \cdot g_2 = g^3 \cdot g_3 = 1$ also $g^1 \cdot g_r = 0$ $r = 2,3$ $g^2 \cdot g_r = 0$ $r = 1,3$ and $g^3 \cdot g_r = 0$ $r = 1,2$, then we have:

$\vec{u},_i \cdot \vec{u},_j = u^1|_i u_1|_j + u^2|_i u_2|_j + u^3|_i u_3|_j$ which can be written in compact form, that is,

$$\vec{u}_{,i} \cdot \vec{u}_{,j} = u^k\big|_i u_k\big|_j \tag{2.1.12d}$$

Substituting Eq. (2.1.12a) and subsequently Eqs. (2.1.12c) and (2.1.12d) into Eq. (2.1.10) leads to:

$$\varepsilon_{ij} = \frac{1}{2}\left(u_i\big|_j + u_j\big|_i + u^k\big|_i u_k\big|_j\right) \tag{2.1.13}$$

Similarly by substituting Eq. (2.1.12b) into Eq. (2.1.11) a term like $G_i \cdot U_m\big|_j G^m$ appears and it can be simplified into $G_i \cdot U_m\big|_j G^m = U_i\big|_j$ and also $G_j \cdot U_m\big|_i G^m = U_j\big|_i$. Moreover, the term $\vec{u}_{,i} \cdot \vec{u}_{,j}$ according to Exercise 2.1.2 could be expanded in terms of contra-variant base vectors and the result will be $\vec{u}_{,i} \cdot \vec{u}_{,j} = U^k\big|_i U_k\big|_j$ and the subsequent substitution of these into Eq. (2.1.11) leads to another equation for Green strain tensor based on the component of the deformed coordinate, that is,

$$\varepsilon_{ij} = \frac{1}{2}\left(U_i\big|_j + U_j\big|_i - U^k\big|_i U_k\big|_j\right) \tag{2.1.14}$$

2.2 STRAIN AND LARGE DEFORMATION IN ORTHOGONAL COORDINATES

Generally speaking in a large deformation any point P in a body with position vector $\vec{r}(X_1, X_2, X_3)$ will be moved to point P' in the deformed body with position vector $\vec{R}(x_1, x_2, x_3)$ and the infinitesimal length dX according to Fig. 2.3, is completely displaced to a new position dx that may be stretched (or contracted) and also rotated relative to its initial position dX. Under this deformation the displacement vector is $d\vec{u}$ and following expressions are obvious:

$$\vec{R} = \vec{r} + \vec{u} \qquad \vec{r} = X_1\vec{i}_1 + X_2\vec{i}_2 + X_3\vec{i}_3 \qquad \vec{R} = x_1\vec{i}_1 + x_2\vec{i}_2 + x_3\vec{i}_3$$

The deformed coordinates x_1, x_2, x_3 are functions of the initial coordinates X_1, X_2, X_3, that is,

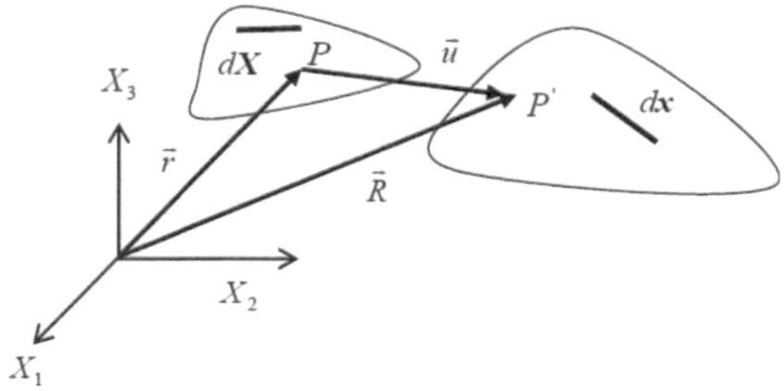

FIGURE 2.3 Un-deformed and deformed position.

$$\begin{aligned} dx_1 &= \frac{\partial x_1}{\partial X_1} dX_1 + \frac{\partial x_1}{\partial X_2} dX_2 + \frac{\partial x_1}{\partial X_3} dX_3 \\ dx_2 &= \frac{\partial x_2}{\partial X_1} dX_1 + \frac{\partial x_2}{\partial X_2} dX_2 + \frac{\partial x_2}{\partial X_3} dX_3 \\ dx_3 &= \frac{\partial x_3}{\partial X_1} dX_1 + \frac{\partial x_3}{\partial X_2} dX_2 + \frac{\partial x_3}{\partial X_3} dX_3 \end{aligned} \tag{2.2.1}$$

The equations in Eq. (2.2.1) can be displayed in matrix form like this:

$$\begin{bmatrix} dx_1 \\ dx_2 \\ dx_3 \end{bmatrix} = \begin{bmatrix} \frac{\partial x_1}{\partial X_1} & \frac{\partial x_1}{\partial X_2} & \frac{\partial x_1}{\partial X_3} \\ \frac{\partial x_2}{\partial X_1} & \frac{\partial x_2}{\partial X_2} & \frac{\partial x_2}{\partial X_3} \\ \frac{\partial x_3}{\partial X_1} & \frac{\partial x_3}{\partial X_2} & \frac{\partial x_3}{\partial X_3} \end{bmatrix} \begin{bmatrix} dX_1 \\ dX_2 \\ dX_3 \end{bmatrix} \tag{2.2.2}$$

The matrix form in Eq. (2) can be expressed in a simple vector and matrix notations like this:

$$\boldsymbol{dx} = \begin{bmatrix} dx_1 \\ dx_2 \\ dx_3 \end{bmatrix} \quad \boldsymbol{dX} = \begin{bmatrix} dX_1 \\ dX_2 \\ dX_3 \end{bmatrix} \quad \boldsymbol{F} = \begin{bmatrix} \frac{\partial x_1}{\partial X_1} & \frac{\partial x_1}{\partial X_2} & \frac{\partial x_1}{\partial X_3} \\ \frac{\partial x_2}{\partial X_1} & \frac{\partial x_2}{\partial X_2} & \frac{\partial x_2}{\partial X_3} \\ \frac{\partial x_3}{\partial X_1} & \frac{\partial x_3}{\partial X_2} & \frac{\partial x_3}{\partial X_3} \end{bmatrix}$$

$$\boldsymbol{dx} = \boldsymbol{F}\boldsymbol{dX} \tag{2.2.3}$$

In Eq. (2.2.3) the matrix $\boldsymbol{F}$ is known as deformation gradient matrix. Since we $x_1 = X_1 + u_1$, $x_2 = X_2 + u_2$ and $x_3 = X_3 + u_3$ such that u_1, u_2, u_3 are components of the displacement vector $\vec{u}$, we can write:

$$\frac{\partial x_i}{\partial X_j} = 1 + \frac{\partial u_i}{\partial X_j} \quad i = j \qquad \frac{\partial x_i}{\partial X_j} = 0 + \frac{\partial u_i}{\partial X_j} \quad i \neq j \tag{2.2.4}$$

According to Eq. (2.2.4) matrix $\boldsymbol{F}$ can be written into following new form:

$$\boldsymbol{F} = \begin{bmatrix} 1+\frac{\partial u_1}{\partial X_1} & \frac{\partial u_1}{\partial X_2} & \frac{\partial u_1}{\partial X_3} \\ \frac{\partial u_2}{\partial X_1} & 1+\frac{\partial u_2}{\partial X_2} & \frac{\partial u_2}{\partial X_3} \\ \frac{\partial u_3}{\partial X_1} & \frac{\partial u_3}{\partial X_2} & 1+\frac{\partial u_3}{\partial X_3} \end{bmatrix} \tag{2.2.5}$$

Expression Eq. (2.2.5) can be shown into simplified form like:

$$\boldsymbol{F} = \underbrace{\boldsymbol{I}}_{3\times 3\, eye\, matrix} + \boldsymbol{D} \quad \boldsymbol{D} = \underbrace{\begin{bmatrix} \frac{\partial u_1}{\partial X_1} & \frac{\partial u_1}{\partial X_2} & \frac{\partial u_1}{\partial X_3} \\ \frac{\partial u_2}{\partial X_1} & \frac{\partial u_2}{\partial X_2} & \frac{\partial u_2}{\partial X_3} \\ \frac{\partial u_3}{\partial X_1} & \frac{\partial u_3}{\partial X_2} & \frac{\partial u_3}{\partial X_3} \end{bmatrix}}_{displacement\ gradient\ matrix} \tag{2.2.6}$$

$\boldsymbol{D}$ is known as displacement gradient matrix, Substituting Eq. (2.2.6) into Eq. (2.2.3) gives an expression in $\boldsymbol{D}$, that is,:

$$\boldsymbol{dx} = (\boldsymbol{I} + \boldsymbol{D})\boldsymbol{dX} \tag{2.2.6}$$

The Green stress tensor can be defined exactly in the same style as 2.1.6, but here the natural coordinates $\alpha_1, \alpha_2, \alpha_3$ are changed to the initial coordinates X_1, X_2, X_3, that is,

$$[\boldsymbol{dx}]^T[\boldsymbol{dx}] - [\boldsymbol{dX}]^T[\boldsymbol{dX}] = 2\,\varepsilon_{ij} dX_i dX_j \tag{2.2.7}$$

However, from (2.2.3) we had:

$$[\boldsymbol{dx}] = [\boldsymbol{F}][\boldsymbol{dX}] \Rightarrow [\boldsymbol{dx}]^T = [\boldsymbol{dX}]^T[\boldsymbol{F}]^T \tag{2.2.8a}$$

Also the quadratic form $\varepsilon_{ij} dX_i dX_j$ can be written into matrix form, that is,

$$\varepsilon_{ij} dX_i dX_j = [\boldsymbol{dX}]^T[\varepsilon][\boldsymbol{dX}] \tag{2.2.8b}$$

Substituting Eqs. (2.2.8a) and (2.2.8b) into Eq. (2.2.7) gives:

$$[\boldsymbol{dX}]^T[\boldsymbol{F}]^T[\boldsymbol{F}][\boldsymbol{dX}] - [\boldsymbol{dX}]^T[\boldsymbol{I}][\boldsymbol{dX}] = 2[\boldsymbol{dX}]^T[\varepsilon][\boldsymbol{dX}]$$

The above expression can be easily simplified to this:

$$[\varepsilon] = \frac{1}{2}\left([\boldsymbol{F}]^T[\boldsymbol{F}] - [\boldsymbol{I}]\right) \tag{2.2.9}$$

Which is called Green strain tensor, it is similar with the definition in Eq. (2.1.7). From Eq. (2.2.6) if we substitute $\boldsymbol{F} = (\boldsymbol{I} + \boldsymbol{D})$ into Eq. (2.2.9) then we have:

$$[\varepsilon] = \frac{1}{2}\left([\boldsymbol{I} + \boldsymbol{D}]^T[\boldsymbol{I} + \boldsymbol{D}] - [\boldsymbol{I}]\right)$$

The above expression can be simplified to:

$$[\varepsilon] = \frac{1}{2}\left(\boldsymbol{D} + \boldsymbol{D}^T + \boldsymbol{D}^T\boldsymbol{D}\right) \tag{2.2.10}$$

If we substitute $\boldsymbol{D}$ from Eq. (2.2.6) into Eq. (2.2.10) then we have:

$$[\varepsilon]=\frac{1}{2}\left(\begin{bmatrix}\frac{\partial u_1}{\partial X_1} & \frac{\partial u_1}{\partial X_2} & \frac{\partial u_1}{\partial X_3}\\ \frac{\partial u_2}{\partial X_1} & \frac{\partial u_2}{\partial X_2} & \frac{\partial u_2}{\partial X_3}\\ \frac{\partial u_3}{\partial X_1} & \frac{\partial u_3}{\partial X_2} & \frac{\partial u_3}{\partial X_3}\end{bmatrix}+\begin{bmatrix}\frac{\partial u_1}{\partial X_1} & \frac{\partial u_2}{\partial X_1} & \frac{\partial u_3}{\partial X_1}\\ \frac{\partial u_1}{\partial X_2} & \frac{\partial u_2}{\partial X_2} & \frac{\partial u_3}{\partial X_2}\\ \frac{\partial u_1}{\partial X_3} & \frac{\partial u_2}{\partial X_3} & \frac{\partial u_3}{\partial X_3}\end{bmatrix}+\begin{bmatrix}\frac{\partial u_1}{\partial X_1} & \frac{\partial u_2}{\partial X_1} & \frac{\partial u_3}{\partial X_1}\\ \frac{\partial u_1}{\partial X_2} & \frac{\partial u_2}{\partial X_2} & \frac{\partial u_3}{\partial X_2}\\ \frac{\partial u_1}{\partial X_3} & \frac{\partial u_2}{\partial X_3} & \frac{\partial u_3}{\partial X_3}\end{bmatrix}\begin{bmatrix}\frac{\partial u_1}{\partial X_1} & \frac{\partial u_1}{\partial X_2} & \frac{\partial u_1}{\partial X_3}\\ \frac{\partial u_2}{\partial X_1} & \frac{\partial u_2}{\partial X_2} & \frac{\partial u_2}{\partial X_3}\\ \frac{\partial u_3}{\partial X_1} & \frac{\partial u_3}{\partial X_2} & \frac{\partial u_3}{\partial X_3}\end{bmatrix}\right)$$

The result is a 3×3 matrix obviously with nine components but it can be reduced to six components because the strain matrix is symmetric which can also be verified from the above expression. We rearrange the components of the strain tensor in a vector into this form:

$$[\varepsilon]=\begin{bmatrix}\varepsilon_{11}\\ \varepsilon_{22}\\ \varepsilon_{33}\\ \varepsilon_{12}\\ \varepsilon_{13}\\ \varepsilon_{23}\end{bmatrix}=\underbrace{\begin{bmatrix}\frac{\partial u_1}{\partial X_1}\\ \frac{\partial u_2}{\partial X_2}\\ \frac{\partial u_3}{\partial X_3}\\ \frac{1}{2}\left(\frac{\partial u_2}{\partial X_1}+\frac{\partial u_1}{\partial X_2}\right)\\ \frac{1}{2}\left(\frac{\partial u_1}{\partial X_3}+\frac{\partial u_3}{\partial X_1}\right)\\ \frac{1}{2}\left(\frac{\partial u_2}{\partial X_3}+\frac{\partial u_3}{\partial X_2}\right)\end{bmatrix}}_{\text{linear part small strain}}+\underbrace{\begin{bmatrix}\frac{1}{2}\left(\left(\frac{\partial u_1}{\partial X_1}\right)^2+\left(\frac{\partial u_2}{\partial X_1}\right)^2+\left(\frac{\partial u_3}{\partial X_1}\right)^2\right)\\ \frac{1}{2}\left(\left(\frac{\partial u_1}{\partial X_2}\right)^2+\left(\frac{\partial u_2}{\partial X_2}\right)^2+\left(\frac{\partial u_3}{\partial X_2}\right)^2\right)\\ \frac{1}{2}\left(\left(\frac{\partial u_1}{\partial X_3}\right)^2+\left(\frac{\partial u_2}{\partial X_3}\right)^2+\left(\frac{\partial u_3}{\partial X_3}\right)^2\right)\\ \frac{1}{2}\left(\frac{\partial u_1}{\partial X_1}\frac{\partial u_1}{\partial X_2}+\frac{\partial u_2}{\partial X_1}\frac{\partial u_2}{\partial X_2}+\frac{\partial u_3}{\partial X_1}\frac{\partial u_3}{\partial X_2}\right)\\ \frac{1}{2}\left(\frac{\partial u_1}{\partial X_1}\frac{\partial u_1}{\partial X_3}+\frac{\partial u_2}{\partial X_1}\frac{\partial u_2}{\partial X_3}+\frac{\partial u_3}{\partial X_1}\frac{\partial u_3}{\partial X_3}\right)\\ \frac{1}{2}\left(\frac{\partial u_1}{\partial X_2}\frac{\partial u_1}{\partial X_3}+\frac{\partial u_2}{\partial X_2}\frac{\partial u_2}{\partial X_3}+\frac{\partial u_3}{\partial X_2}\frac{\partial u_3}{\partial X_3}\right)\end{bmatrix}}_{\text{nonlinear part large strain}} \tag{2.2.11}$$

It should be reminded that if the deformation is pure rotation then the strain tensor vanishes, because $\boldsymbol{dx} = \boldsymbol{R}\,\boldsymbol{dX}$ where $\boldsymbol{R}$ or rotation matrix is an orthogonal one, that is, $\boldsymbol{R}\,\boldsymbol{R}^T = \boldsymbol{I}$, in this case when $\boldsymbol{R} = \boldsymbol{R}$, then $\boldsymbol{D} = \boldsymbol{R} - \boldsymbol{I}$ and if we substitute this in (10) then we have:

$$[\varepsilon] = \frac{1}{2}\left([\boldsymbol{I}+\boldsymbol{D}]^T[\boldsymbol{I}+\boldsymbol{D}] - [\boldsymbol{I}]\right) = \frac{1}{2}\left([\boldsymbol{R}]^T[\boldsymbol{R}] - [\boldsymbol{I}]\right) = [\boldsymbol{0}]$$

In another special case when deformation small term $\boldsymbol{D}^T\boldsymbol{D}$ in (2.2.10) or the 2nd bracket in Eq. (2.2.11) can be ignored in this case the strain tensor will change to:

$$[\varepsilon] \cong \frac{1}{2}\left(\boldsymbol{D} + \boldsymbol{D}^T\right) \tag{2.2.10a}$$

Now can compare the strain formulas in this section (for orthogonal coordinate), with the ones derived in previous section (for natural coordinate). If we interpret these formulas in terms of natural coordinate then it is necessary to demonstrate that the end result is the formula for Green strain tensor given by Eq. (2.1.7). In Chapter 1, we mentioned that $\vec{r} = X_1\vec{I}_1 + X_2\vec{I}_2 + X_3\vec{I}_3$, but here it should be replaced with $\boldsymbol{X} = X_1\vec{I}_1 + X_2\vec{I}_2 + X_3\vec{I}_3$, that is, this change $\vec{r} \Leftrightarrow X$ should be made, and then the Eq. (1.2.7) can be written as:

$$\boldsymbol{dX} = \frac{\partial \boldsymbol{X}}{\partial \alpha_i} d\alpha^i = g_i\, d\alpha^i \quad i = 1,2,3 \tag{2.2.12}$$

In Eq. (2.2.12) α_1, α_2 and α_3 are the natural coordinates and g_1, g_2 and g_3 are covariant base vectors, that can be expressed by:

$$g_i = \frac{\partial X_1}{\partial \alpha_i}\vec{I}_1 + \frac{\partial X_2}{\partial \alpha_i}\vec{I}_2 + \frac{\partial X_3}{\partial \alpha_i}\vec{I}_3 \quad i = 1,2,3 \tag{2.2.12a}$$

In the deformed body $\vec{R} = x_1\vec{I}_1 + x_2\vec{I}_2 + x_3\vec{I}_3$ should be replaced with $\boldsymbol{x} = x_1\vec{I}_1 + x_2\vec{I}_2 + x_3\vec{I}_3$, that is, this change $\vec{R} \Leftrightarrow x$ should be made, and then the Eq. (2.1.7) can be written as:

$$\boldsymbol{dx} = \frac{\partial \boldsymbol{x}}{\partial \alpha_i} d\alpha^i = G_i\, d\alpha^i \tag{2.2.13}$$

In Eq. (2.2.13) G_1, G_2 and G_3 are deformed covariant base vectors, that can be expressed by:

$$G_i = \frac{\partial x_1}{\partial \alpha_i}\vec{I}_1 + \frac{\partial x_2}{\partial \alpha_i}\vec{I}_2 + \frac{\partial x_3}{\partial \alpha_i}\vec{I}_3 \qquad i = 1,2,3 \tag{2.2.13a}$$

Deformation gradient matrix was defined like this:

$$\boldsymbol{dx} = \boldsymbol{F}\,\boldsymbol{dX} \tag{2.2.14}$$

By substituting Eqs. (2.2.12) and (2.2.13) into Eq. (2.2.14) we have:

$$G_i\,d\alpha^i = \boldsymbol{F}\,g_i\,d\alpha^i \Rightarrow G_i = \boldsymbol{F}\,g_i \qquad i = 1,2,3 \tag{2.2.15}$$

In vector-matrix equation (2.2.15) G_i and g_i are also vectors but not shown by bold characters only because they are base vectors.

Exercise 2.2.1: *Expand matrix equation* (2.2.15)

Solution: In fact each of them have three components and described by Eqs. (2.2.12a) and (2.2.13a) and if we substitute it in Eq. (2.2.15) together with definition of ***F*** from Eq. (2.2.5) it results three expressions (each one for a component) such as:

$$\begin{aligned}
\frac{\partial x_1}{\partial \alpha_i} &= \frac{\partial x_1}{\partial X_1}\frac{\partial X_1}{\partial \alpha_i} + \frac{\partial x_1}{\partial X_2}\frac{\partial X_2}{\partial \alpha_i} + \frac{\partial x_1}{\partial X_3}\frac{\partial X_3}{\partial \alpha_i} \qquad i = 1,2,3 \\
\frac{\partial x_2}{\partial \alpha_i} &= \frac{\partial x_2}{\partial X_1}\frac{\partial X_1}{\partial \alpha_i} + \frac{\partial x_2}{\partial X_2}\frac{\partial X_2}{\partial \alpha_i} + \frac{\partial x_2}{\partial X_3}\frac{\partial X_3}{\partial \alpha_i} \qquad i = 1,2,3 \\
\frac{\partial x_3}{\partial \alpha_i} &= \frac{\partial x_3}{\partial X_1}\frac{\partial X_1}{\partial \alpha_i} + \frac{\partial x_3}{\partial X_2}\frac{\partial X_2}{\partial \alpha_i} + \frac{\partial x_3}{\partial X_3}\frac{\partial X_3}{\partial \alpha_i} \qquad i = 1,2,3
\end{aligned} \tag{2.2.15a}$$

It is obvious that the set of nine equations in Eq. (2.2.15a) and nine equations in Eq. (2.2.15) are identical. The former represent a chain rule of differentiation applied to the components, while the latter expresses the relationship between deformation gradient tensor $\boldsymbol{F}$ with the metric tensor. To find this we need to post multiply Eq. (2.2.15) into g^i, this will be matrix (outer) product and not scalar (inner) product, then we have:

$$G_i \otimes g^i = \boldsymbol{F}\, g_i \otimes g^i$$

Expanding the above expression provides this form:

$$G_1 \otimes g^1 + G_2 \otimes g^2 + G_3 \otimes g^3 = \boldsymbol{F}\begin{bmatrix} g_1 \otimes g^1 + g_2 \otimes g^2 + \\ g_3 \otimes g^3 \end{bmatrix} \qquad (2.2.15b)$$

In the above expression we expand the bracket we have:

$$g_1 \otimes g^1 + g_2 \otimes g^2 + g_3 \otimes g^3 = \underbrace{\begin{bmatrix} 1 & 0 & 0 \\ 0 & 0 & 0 \\ 0 & 0 & 0 \end{bmatrix}}_{g_1 \otimes g^1} + \underbrace{\begin{bmatrix} 0 & 0 & 0 \\ 0 & 1 & 0 \\ 0 & 0 & 0 \end{bmatrix}}_{g_2 \otimes g^2} + \underbrace{\begin{bmatrix} 1 & 0 & 0 \\ 0 & 0 & 0 \\ 0 & 0 & 0 \end{bmatrix}}_{g_3 \otimes g^3} = \boldsymbol{I}$$

Substituting the above into Eq. (2.2.15b) then we find $\boldsymbol{F}$ versus:

$$\boldsymbol{F} = G_1 \otimes g^1 + G_2 \otimes g^2 + G_3 \otimes g^3 = [G_i][g^i]^T \qquad (2.2.16)$$

By transposing (2.2.16) we have:

$$\boldsymbol{F}^T = [g^j][G_j]^T \qquad (2.2.17)$$

Now substitute both Eqs. (2.2.16) and (2.2.16) into Eq. (2.2.9) to find the strain tensor, that is,

$$[\varepsilon] = \frac{1}{2}\begin{pmatrix} [g^j][G_j]^T \\ [G_i][g^i]^T - \boldsymbol{I} \end{pmatrix} = \frac{1}{2}\begin{pmatrix} [g^j][G_j]^T[G_i][g^i]^T \\ -\underbrace{[g_i][g^i]^T}_{\boldsymbol{I}} \end{pmatrix} \qquad (2.2.18)$$

By post multiplying Eq. (2.2.18) into g_i considering that $\left[g^i\right]^T[g_i]=1$ then:

$$[\varepsilon][g_i]=\frac{1}{2}\begin{pmatrix}[g^j][G_j]^T[G_i]\underbrace{\left[g^i\right]^T[g_i]}_{1}\\ -\boldsymbol{I}[g_i]\end{pmatrix}=\frac{1}{2}\begin{pmatrix}[g^j][G_j]^T[G_i]\\ -[g_i]\end{pmatrix} \tag{2.2.19}$$

The Eq. (2.2.19) now is taken the form of a column vector $[\varepsilon][g_i]$ and if we want to change it to a scalar type it is necessary to pre multiply both sides into $\left[g_j\right]^T$ also considering that $\left[g_j\right]^T\left[g^j\right]=1$, then we have:

$$\left[g_j\right]^T[\varepsilon][g_i]=\frac{1}{2}\left(\underbrace{\left[g_j\right]^T\left[g^j\right]}_{1}\left[G_j\right]^T[G_i]-\left[g_j\right]^T[g_i]\right) \tag{2.2.19a}$$

Exercise 2.2.2: *Simplify all the terms* in Eq. (2.2.19a)

Solution: I chapter I, we showed that $\left[g_j\right]^T[g_i]=g_i\cdot g_j=g_{ij}$ and $\left[G_j\right]^T[G_i]=G_i\cdot G_j=G_{ij}$, also $\left[g_j\right]^T[\varepsilon][g_i]$ is an scalar like the other terms of Eq. (2.2.19a). In order to prove consider for example $j=2$ and $i=1$, then we have:

$$[g_2]^T[\varepsilon][g_1]=\begin{bmatrix}0 & 1 & 0\end{bmatrix}\begin{bmatrix}\varepsilon_{11} & \varepsilon_{12} & \varepsilon_{13}\\ \varepsilon_{21} & \varepsilon_{22} & \varepsilon_{23}\\ \varepsilon_{31} & \varepsilon_{32} & \varepsilon_{33}\end{bmatrix}\begin{bmatrix}1\\0\\0\end{bmatrix}=\begin{bmatrix}0 & 1 & 0\end{bmatrix}\begin{bmatrix}\varepsilon_{11}\\ \varepsilon_{21}\\ \varepsilon_{31}\end{bmatrix}=\varepsilon_{21}$$

The above example shows that $\left[g_j\right]^T[\varepsilon][g_i]=g_{ji}=g_{ij}$ and if we substitute all in (2.2.19a) then we have:

$$\varepsilon_{ij}=\frac{1}{2}\left(G_{ij}-g_{ij}\right) \tag{2.2.19b}$$

The above expression is the Green strain tensor we introduced in last section and is also derivable from Eq. (2.2.9) as we demonstrated above.

We can also demonstrate the reverse procedure, that is, from Eq. (2.2.19b) above the Eq. (2.2.9) is also derivable, just we need to replace the natural coordinates α_1, α_2 and α_3 with orthogonal coordinates X_1, X_2 and X_3 which also can be shown in body attached frames, that is, $\vec{r} = X_1 \vec{i}_1 + X_2 \vec{i}_2 + X_3 \vec{i}_3$ and $\vec{R} = x_1 \vec{i}_1 + x_2 \vec{i}_2 + x_3 \vec{i}_3$. The covariant base vectors are $g_1 = \vec{i}_1$, $g_2 = \vec{i}_2$ and $g_3 = \vec{i}_3$ in general we can write:

$$g_i = \frac{\partial \vec{r}}{\partial X_i} \Rightarrow g_{ij} = \delta_{ij}$$

The above expression was also discussed in chapter one. However, for the deformed body with position vector $\vec{R}$ the covariant base vectors are:

$$G_i = \frac{\partial \vec{R}}{\partial X_i} = \frac{\partial x_1}{\partial X_i} \vec{i}_1 + \frac{\partial x_2}{\partial X_i} \vec{i}_2 + \frac{\partial x_3}{\partial X_i} \vec{i}_3$$

$$G_j = \frac{\partial \vec{R}}{\partial X_j} = \frac{\partial x_1}{\partial X_j} \vec{i}_1 + \frac{\partial x_2}{\partial X_j} \vec{i}_2 + \frac{\partial x_3}{\partial X_j} \vec{i}_3$$

Also we have: $G_{ij} = G_i \cdot G_j$

Substituting the expressions in G_{ij} we have:

$$G_{ij} = \left(\frac{\partial x_1}{\partial X_i} \vec{i}_1 + \frac{\partial x_2}{\partial X_i} \vec{i}_2 + \frac{\partial x_3}{\partial X_i} \vec{i}_3 \right) \cdot \left(\frac{\partial x_1}{\partial X_j} \vec{i}_1 + \frac{\partial x_2}{\partial X_j} \vec{i}_2 + \frac{\partial x_3}{\partial X_j} \vec{i}_3 \right)$$

The result of the above is:

$$G_{ij} = \frac{\partial x_1}{\partial X_i} \frac{\partial x_1}{\partial X_j} + \frac{\partial x_2}{\partial X_i} \frac{\partial x_2}{\partial X_j} + \frac{\partial x_3}{\partial X_i} \frac{\partial x_3}{\partial X_j}$$

The above expression can simplified by summation index, that is,

$$G_{ij} = \frac{\partial x_k}{\partial X_i} \frac{\partial x_k}{\partial X_j} \qquad i, j = 1, 2, 3$$

Now substituting G_{ij} and $g_{ij} = \delta_{ij}$ into Green strain tensor formula (2.2.19b) then:

$$\varepsilon_{ij} = \frac{1}{2}\left(\frac{\partial x_k}{\partial X_i}\frac{\partial x_k}{\partial X_j} - \delta_{ij}\right)$$

The term $\frac{\partial x_k}{\partial X_i}$ are the rows of the matrix $\boldsymbol{F}^T$ also term $\frac{\partial x_k}{\partial X_j}$ are the rows of the matrix $\boldsymbol{F}$, therefore the above expression can be written into matrix form, that is,

$$[\varepsilon] = \frac{1}{2}\left(\boldsymbol{F}^T\boldsymbol{F} - \boldsymbol{I}\right)$$

That is identical with Eq. (2.2.9) so the proof is complete.

2.3 DECOMPOSITION OF DEFORMATION GRADIENT MATRIX

In previous section, it was demonstrated that large deformations and strains in orthogonal coordinate system can be expressed by deformation gradient matrix. The advantage of using this matrix lies on its decomposition into two components each represent a distinct property of the large deformation. It already known that (see Eq. (2.2.3)):

$$\boldsymbol{dx} = \frac{\partial \boldsymbol{x}}{\partial \boldsymbol{X}}\boldsymbol{dX} = \boldsymbol{F}\,\boldsymbol{dX} \tag{2.3.1}$$

The equation (2.3.1) can be expanded like this:

$$\boldsymbol{dx} = \frac{\partial \boldsymbol{x}}{\partial \boldsymbol{x}_r}\frac{\partial \boldsymbol{x}_r}{\partial \boldsymbol{X}}\boldsymbol{dX} \tag{2.3.2}$$

In the above expansion $\frac{\partial \boldsymbol{x}}{\partial \boldsymbol{x}_r}$ and $\frac{\partial \boldsymbol{x}_r}{\partial \boldsymbol{X}}$ each expresses a matrix individually. It looks like that the main matrix $\boldsymbol{F}$ is decomposed into two matrices that their matrix product results $\boldsymbol{F}$, the vector $\boldsymbol{x}_r$ represents an intermediate coordinate such that:

$$\frac{\partial \boldsymbol{x}}{\partial \boldsymbol{x}_r} = \boldsymbol{R} \tag{2.3.3}$$

In Eq. (2.3.3), the matrix $\boldsymbol{R}$ expresses pure rotation such that $\boldsymbol{R}\boldsymbol{R}^T = \boldsymbol{I}$, by pure rotation we mean rigid body type rotation. This is called "polar decomposition" in which rigid body rotation $\boldsymbol{R}$ is kept separate. Therefore, pure deformation part $\boldsymbol{U}_R$ can be defined by:

$$\frac{\partial \boldsymbol{x}_r}{\partial \boldsymbol{X}} = \boldsymbol{U}_R \tag{2.3.4}$$

The matrix $\boldsymbol{U}_R$ that is defined by Eq. (2.3.4) is called "right stretch tensor" and if we substitute Eqs. (2.3.3) and (2.3.4) in Eq. (2.3.2) and then compare the result with Eq. (2.3.1) we have:

$$\boldsymbol{F} = \boldsymbol{R}\,\boldsymbol{U}_R \tag{2.3.5a}$$

By changing the order of multiplication and following the same procedure we can define another matrix $\boldsymbol{V}$ called "left stretch tensor" and in general a following statement can be made:

Through using polar decomposition, the deformation gradient tensor $\boldsymbol{F}$ can be decomposed to rotation tensor $\boldsymbol{R}$ either with right stretch tensor $\boldsymbol{U}_R$ or left stretch tensor $\boldsymbol{V}$ and the general formula is:

$$\boldsymbol{F} = \boldsymbol{R}\,\boldsymbol{U}_R = \boldsymbol{V}\,\boldsymbol{R} \tag{2.3.5b}$$

In order to find what the stretch tensors $\boldsymbol{U}_R$ (from now $\boldsymbol{U}$) and $\boldsymbol{V}$ are? First we define engineering concept of stretch which is called λ and is a number greater than 1, $\lambda > 1$ and can be expressed by:

$$\lambda = \frac{dl}{dl_0} \qquad \lambda^2 = \frac{dl^2}{dl_0^2} \tag{2.3.6}$$

In Eq. (2.3.6) the infinitesimal length dl is the deformed length and can be written in terms of the vector $\boldsymbol{dx}$ as follows:

$$dl^2 = [\boldsymbol{dx}]^T[\boldsymbol{dx}] = dx_1^2 + dx_2^2 + dx_3^2 \tag{2.3.7}$$

Similarly infinitesimal length $d\,l_0$ is the initial (un-deformed) length and can be written in terms of the vector $\boldsymbol{dX}$ as follows:

$$d\,l_0^2 = [\boldsymbol{dX}]^T [\boldsymbol{dX}] = dX_1^2 + dX_2^2 + dX_3^2 \tag{2.3.8}$$

If we show the unit vector in the $\boldsymbol{dX}$ direction by $\boldsymbol{N}$, then we have:

$$\boldsymbol{N} = \frac{\boldsymbol{dX}}{\|\boldsymbol{dX}\|} = \frac{\boldsymbol{dX}}{\left(\boldsymbol{dX}^T\boldsymbol{dX}\right)^{1/2}} \Rightarrow \boldsymbol{N}^T = \frac{\boldsymbol{dX}^T}{\left(\boldsymbol{dX}^T\boldsymbol{dX}\right)^{1/2}} \tag{2.3.9}$$

Meanwhile from Eq. (2.3.1) we can write:

$$\boldsymbol{dx} = \boldsymbol{F}\,\boldsymbol{dX} \Rightarrow \boldsymbol{dx}^T = \boldsymbol{dX}^T\boldsymbol{F}^T \tag{2.3.10}$$

Exercise 2.3.1: *Find a formula* for λ^2

Solution: If we substitute Eqs. (2.3.7) and (2.3.8) into Eq. (2.3.6) then we have:

$$\lambda^2 = \frac{d\,l^2}{d\,l_0^2} = \frac{\boldsymbol{dx}^T\boldsymbol{dx}}{\boldsymbol{dX}^T\boldsymbol{dX}} \tag{2.3.10a}$$

Substituting Eq. (2.3.9) into Eq. (2.3.10), results new form of Eq. (2.3.10), which is:

$$\boldsymbol{dx} = \boldsymbol{FN}\left(\boldsymbol{dX}^T\boldsymbol{dX}\right)^{1/2} \Rightarrow \boldsymbol{dx}^T = \boldsymbol{N}^T\boldsymbol{F}^T\left(\boldsymbol{dX}^T\boldsymbol{dX}\right)^{1/2} \tag{2.3.10b}$$

Substituting Eq. (2.3.10b) into Eq. (2.3.10a) gives:

$$\lambda^2 = \frac{d\,l^2}{d\,l_0^2} = \frac{\boldsymbol{N}^T\boldsymbol{F}^T\left(\boldsymbol{dX}^T\boldsymbol{dX}\right)^{1/2}\boldsymbol{FN}\left(\boldsymbol{dX}^T\boldsymbol{dX}\right)^{1/2}}{\boldsymbol{dX}^T\boldsymbol{dX}}$$

This can be simplified into:

$$\lambda^2 = \boldsymbol{N}^T\boldsymbol{F}^T\boldsymbol{F}\,\boldsymbol{N} \tag{2.3.11}$$

Any unit vector has the length 1, therefore we have:

$$\boldsymbol{N}^T \boldsymbol{N} - 1 = 0 \tag{2.3.12}$$

In Eq. (2.3.11), λ^2 is square of the stretch value and it depends on the direction $\boldsymbol{N}$. In order to keep the variation of λ^2 minimum or $\delta\left(\lambda^2\right)=0$ when the $\boldsymbol{N}$ is changing and also we have a constraint (2.3.12) to satisfy we use Lagrange multiplier method. In this method we define a multiplier called η and specify the Lagrange function $\phi=\delta\left(\lambda^2\right)$ like this:

$$\phi = \boldsymbol{N}^T \boldsymbol{F}^T \boldsymbol{F}\boldsymbol{N} + \eta\left(\boldsymbol{N}^T \boldsymbol{N} - 1\right) \tag{2.3.13a}$$

In order to find a direction in which λ^2 remains stationary or $\delta\left(\lambda^2\right)=0$, we need to use the Lagrange rule that can be written as follows:

$$\frac{\partial \phi}{\partial \boldsymbol{N}^T} = \boldsymbol{F}^T \boldsymbol{F}\, \boldsymbol{N} - \eta\, \boldsymbol{I}\, \boldsymbol{N} = 0 \tag{2.3.13b}$$

When we post-multiply both sides of the above vector expression into $\boldsymbol{N}^T$, then we have:

$$\frac{\partial \phi}{\partial \boldsymbol{N}^T} = \boldsymbol{N}^T \boldsymbol{F}^T \boldsymbol{F}\, \boldsymbol{N} - \eta\, \underbrace{\boldsymbol{N}^T\, \boldsymbol{I}\, \boldsymbol{N}}_{1} = 0$$

The above expression results:

$$\eta = \boldsymbol{N}^T \boldsymbol{F}^T \boldsymbol{F}\, \boldsymbol{N} \tag{2.3.14}$$

Comparing Eqs. (2.3.11) and (2.3.14) shows that Lagrange multiplier η, is the same as square of stretch λ^2, that is, $\eta=\lambda^2$. Therefore, in Eq. (2.3.13b) instead of $\boldsymbol{F}^T \boldsymbol{F}\, \boldsymbol{N} - \eta\, \boldsymbol{I}\, \boldsymbol{N} = 0$ we can write:

$$\left(\boldsymbol{F}^T \boldsymbol{F} - \lambda^2\, \boldsymbol{I}\right)\boldsymbol{N} = \left(\boldsymbol{b} - \lambda^2\, \boldsymbol{I}\right)\boldsymbol{N} = 0 \tag{2.3.15}$$

In Eq. (2.3.15) the tensor $\boldsymbol{b} = \boldsymbol{F}^T \boldsymbol{F}$ is called “right Cauchy Green tensor.” According to Eq. (2.3.15) the square of stretches λ^2 are the eigenvalues of the right Cauchy Green tensor $\boldsymbol{b}$. Meanwhile by help of Eq. (2.3.5a) we can write:

$$\boldsymbol{F}^T\boldsymbol{F} = \boldsymbol{U}^T\boldsymbol{R}^T\boldsymbol{R}\boldsymbol{U}$$

Since $\boldsymbol{R}^T\boldsymbol{R} = \boldsymbol{I}$, the above expression can be written into this simplified form:

$$\boldsymbol{F}^T\boldsymbol{F} = \boldsymbol{U}^T\boldsymbol{U} \tag{2.3.16}$$

Substituting Eq. (2.3.16) into Eq. (2.3.15) provides new expression such as:

$$\left(\boldsymbol{U}^T\boldsymbol{U} - \lambda^2\boldsymbol{I}\right)\boldsymbol{N} = 0 \tag{2.3.17}$$

If we name the eigenvectors matrices $\boldsymbol{Q}(\boldsymbol{N}) = [\boldsymbol{N}_1 \quad \boldsymbol{N}_2 \quad \boldsymbol{N}_3]$, in which every Eigen vector forms a column in $\boldsymbol{Q}(\boldsymbol{N})$, then according to a theorem in linear algebra we have:

$$\boldsymbol{U}^T\boldsymbol{U} = \boldsymbol{Q}(\boldsymbol{N})\operatorname{diag}\left(\lambda_1^2, \lambda_2^2, \lambda_3^2\right)\boldsymbol{Q}(\boldsymbol{N})^T \tag{2.3.18}$$

Exercise 2.3.2: *Prove the expression* in Eq. (2.3.18)

Proof: The Eq. (2.3.17) can be written into this form:

$$\left(\boldsymbol{U}^T\boldsymbol{U} - \lambda_i^2\boldsymbol{I}\right)\boldsymbol{N}_i = 0 \quad i = 1,2,3$$

This results,

$$\boldsymbol{U}^T\boldsymbol{U}\boldsymbol{N}_i = \lambda_i^2\boldsymbol{I}\boldsymbol{N}_i = \lambda_i^2\boldsymbol{N}_i \quad i = 1,2,3 \tag{2.3.18a}$$

In Eq. (2.3.18a) each $\boldsymbol{N}_i \quad i = 1,2,3$ is a column vector the whole (18a) represents 3 column vectors identity. If we use definition of $\boldsymbol{Q}(\boldsymbol{N}) = [\boldsymbol{N}_1 \quad \boldsymbol{N}_2 \quad \boldsymbol{N}_3]$, then the Eq. (2.3.18a) can be transformed into a matrix form like this:

$$\boldsymbol{U}^T\boldsymbol{U}\boldsymbol{Q}(\boldsymbol{N}) = \boldsymbol{Q}(\boldsymbol{N})\operatorname{diag}\left(\lambda_1^2, \lambda_2^2, \lambda_3^2\right) \tag{2.3.18b}$$

By post-multiplying both sides of Eq. (2.3.18b) into $\boldsymbol{Q}(\boldsymbol{N})^T = \left[\boldsymbol{N}_1^T \quad \boldsymbol{N}_2^T \quad \boldsymbol{N}_3^T\right]$ where $\boldsymbol{N}_1^T$, $\boldsymbol{N}_2^T$ and $\boldsymbol{N}_3^T$ are the row vectors we have:

$$\boldsymbol{U}^T \boldsymbol{U} \boldsymbol{Q}(\boldsymbol{N}) \boldsymbol{Q}(\boldsymbol{N})^T = \boldsymbol{Q}(\boldsymbol{N}) \operatorname{diag}\left(\lambda_1^2, \lambda_2^2, \lambda_3^2\right) \boldsymbol{Q}(\boldsymbol{N})^T$$

Since eigenvector matrices are always normalized, then when can write $\boldsymbol{Q}(\boldsymbol{N})\boldsymbol{Q}(\boldsymbol{N})^T = \boldsymbol{I}$ and the above expression will change to:

$$\boldsymbol{U}^T \boldsymbol{U} = \boldsymbol{Q}(\boldsymbol{N}) \operatorname{diag}\left(\lambda_1^2, \lambda_2^2, \lambda_3^2\right) \boldsymbol{Q}(\boldsymbol{N})^T$$

Which is identical to Eq. (2.3.18) and the proof is complete.

If we want to expand Eq. (2.3.18) to find out the building blocks of $\boldsymbol{U}^T \boldsymbol{U}$ we have:

$$\boldsymbol{U}^T \boldsymbol{U} = \begin{bmatrix} n_{11} & n_{21} & n_{31} \\ n_{12} & n_{22} & n_{32} \\ n_{13} & n_{23} & n_{33} \end{bmatrix} \begin{bmatrix} \lambda_1^2 & 0 & 0 \\ 0 & \lambda_2^2 & 0 \\ 0 & 0 & \lambda_3^2 \end{bmatrix} \begin{bmatrix} n_{11} & n_{12} & n_{13} \\ n_{21} & n_{22} & n_{23} \\ n_{31} & n_{32} & n_{33} \end{bmatrix}$$

$$\boldsymbol{N}_i = \begin{bmatrix} n_{i1} \\ n_{i2} \\ n_{i3} \end{bmatrix} \quad \boldsymbol{N}_i^T = \begin{bmatrix} n_{i1} & n_{i2} & n_{i3} \end{bmatrix} \quad i = 1,2,3$$

The above expression can be simplified in terms of Eigen vectors, that is,

$$\boldsymbol{U}^T \boldsymbol{U} = \begin{bmatrix} \lambda_1^2 \boldsymbol{N}_1 & \lambda_2^2 \boldsymbol{N}_2 & \lambda_3^2 \boldsymbol{N}_3 \end{bmatrix} \begin{bmatrix} \boldsymbol{N}_1^T \\ \boldsymbol{N}_2^T \\ \boldsymbol{N}_3^T \end{bmatrix} \tag{2.3.19a}$$

The expansion of the above yields to this:

$$\boldsymbol{U}^T \boldsymbol{U} = \lambda_1^2 \boldsymbol{N}_1 \boldsymbol{N}_1^T + \lambda_2^2 \boldsymbol{N}_2 \boldsymbol{N}_2^T + \lambda_3^2 \boldsymbol{N}_3 \boldsymbol{N}_3^T \tag{2.3.19b}$$

Since the matrix $\boldsymbol{U}$ is symmetric then $\boldsymbol{U}^T = \boldsymbol{U}$ and also $\boldsymbol{U}^T \boldsymbol{U} = \boldsymbol{U}^2$, then from Eq. (2.3.17) we have:

$$\left(\boldsymbol{U}^T \boldsymbol{U} - \lambda^2 \boldsymbol{I}\right)\boldsymbol{N} = \left(\boldsymbol{U}^T + \lambda \boldsymbol{I}\right)\left(\boldsymbol{U} - \lambda \boldsymbol{I}\right)\boldsymbol{N} = 0$$

From the above expression we can conclude that:

$$(\boldsymbol{U}-\lambda\boldsymbol{I})\boldsymbol{N}=0 \tag{2.3.20a}$$

Similar to the previous proof it can also be shown that:

$$\boldsymbol{U}=\lambda_1\boldsymbol{N}_1\boldsymbol{N}_1^T+\lambda_2\boldsymbol{N}_2\boldsymbol{N}_2^T+\lambda_3\boldsymbol{N}_3\boldsymbol{N}_3^T \tag{2.3.20b}$$

Equations Eqs. (2.3.20a) and (2.3.20b) can be interpreted in physical term, expressing that the principal stretches λ_1, λ_2 and λ_3 are the eigenvalues of the right stretch tensor $\boldsymbol{U}$. These stretches are in directions $\boldsymbol{N}_1$, $\boldsymbol{N}_2$ and $\boldsymbol{N}_3$ respectively and they are perpendicular to each other. Therefore, if the matrix $\boldsymbol{F}$ that contains information about the deformation undergoes polar decomposition $\boldsymbol{F}=\boldsymbol{R}\,\boldsymbol{U}$ then from matrix $\boldsymbol{U}$, we can find the principal stretches λ_1, λ_2 and λ_3 together with their directions $\boldsymbol{N}_1$, $\boldsymbol{N}_2$ and $\boldsymbol{N}_3$. Moreover, the theorem (2.3.18) is also valid for $\boldsymbol{U}$, that is,

$$\boldsymbol{U}=\boldsymbol{Q}(\boldsymbol{N})\operatorname{diag}(\lambda_1,\lambda_2,\lambda_3)\boldsymbol{Q}(\boldsymbol{N})^T \tag{2.3.20c}$$

Similarly in the deformed status, the unit vector $\boldsymbol{n}$ in direction $\boldsymbol{dx}$ can be defined as follows (see Eq. (2.3.9)):

$$\boldsymbol{n}=\frac{\boldsymbol{dx}}{\|\boldsymbol{dx}\|}=\frac{\boldsymbol{dx}}{\left(\boldsymbol{dx}^T\boldsymbol{dx}\right)^{1/2}} \qquad \boldsymbol{dx}=\boldsymbol{F}\,\boldsymbol{dX} \tag{2.3.21}$$

Equation (2.3.21) can be expanded by using Eq. (2.3.9), which says $\boldsymbol{dX}=\boldsymbol{N}\|\boldsymbol{dX}\|$ into this form:

$$\boldsymbol{n}=\frac{\boldsymbol{F}\,\boldsymbol{dX}}{\|\boldsymbol{dx}\|}=\frac{\boldsymbol{F}\,\boldsymbol{N}\|\boldsymbol{dX}\|}{\|\boldsymbol{dx}\|}$$

Since $\dfrac{\|\boldsymbol{dx}\|}{\|\boldsymbol{dX}\|}=\lambda$ or stretch value, then $\dfrac{\|\boldsymbol{dX}\|}{\|\boldsymbol{dx}\|}=\dfrac{1}{\lambda}$ and the above equation changes to:

$$\boldsymbol{n}=\frac{1}{\lambda}\boldsymbol{F}\,\boldsymbol{N} \tag{2.3.22}$$

From Eq. (2.3.22) it is obvious that $\boldsymbol{F}\,\boldsymbol{N} = \lambda\,\boldsymbol{n}$ and also $\boldsymbol{N} = \lambda\,\boldsymbol{F}^{-1}\boldsymbol{n}$ and if we substitute both of these into Eq. (2.3.15) or $\boldsymbol{F}^T\boldsymbol{F}\,\boldsymbol{N} - \lambda^2\,\boldsymbol{N} = 0$ we have:

$$\boldsymbol{F}^T\lambda\,\boldsymbol{n} - \lambda^3\,\boldsymbol{F}^{-1}\boldsymbol{n} = 0 \qquad (2.3.23)$$

By pre-multiplying both sides of Eq. (2.3.23) into $\frac{1}{\lambda}\boldsymbol{F}$ we have:

$$\left(\boldsymbol{F}\,\boldsymbol{F}^T - \lambda^2\,\boldsymbol{I}\right)\boldsymbol{n} = 0 \qquad (2.3.24)$$

In Eq. (2.3.24) the matrix $\boldsymbol{c} = \boldsymbol{F}\,\boldsymbol{F}^T$ is called "left Cauchy Green tensor." Then by substituting $\boldsymbol{F} = \boldsymbol{V}\,\boldsymbol{R}$ and $\boldsymbol{F}^T = \boldsymbol{R}^T\boldsymbol{V}^T$ into Eq. (2.3.24) we have:

$$\left(\boldsymbol{V}\boldsymbol{R}\boldsymbol{R}^T\boldsymbol{V}^T - \lambda^2\,\boldsymbol{I}\right)\boldsymbol{n} = 0$$

Since $\boldsymbol{R}\boldsymbol{R}^T = \boldsymbol{I}$ the above equation will change to:

$$\left(\boldsymbol{V}\,\boldsymbol{V}^T - \lambda^2\,\boldsymbol{I}\right)\boldsymbol{n} = 0 \qquad (2.3.25)$$

Equation (2.3.25) is similar to Eq. (2.3.17) and $\boldsymbol{V}$ is the "left stretch tensor" and it is also symmetric, that is, $\boldsymbol{V} = \boldsymbol{V}^T$, therefore equations similar to Eqs. (2.3.20a), (2.3.20b) and (2.3.20c) can be written in this case only the eigenvector matrix changes to $\boldsymbol{q}(\boldsymbol{n}) = [\boldsymbol{n}_1 \quad \boldsymbol{n}_2 \quad \boldsymbol{n}_3]$ and we have:

$$\boldsymbol{V} = \lambda_1\,\boldsymbol{n}_1\boldsymbol{n}_1^T + \lambda_2\,\boldsymbol{n}_2\boldsymbol{n}_2^T + \lambda_3\,\boldsymbol{n}_3\boldsymbol{n}_3^T \qquad (2.3.26)$$

$$\left(\boldsymbol{V} - \lambda\,\boldsymbol{I}\right)\boldsymbol{n} = 0 \qquad (2.3.27a)$$

$$\boldsymbol{V} = \boldsymbol{q}(\boldsymbol{n})\operatorname{diag}(\lambda_1,\lambda_2,\lambda_3)\,\boldsymbol{q}(\boldsymbol{n})^T \qquad (2.3.27b)$$

The unit vectors or eigenvectors $\boldsymbol{n}_1,\boldsymbol{n}_2$ and $\boldsymbol{n}_3$ are called "Euler triads" while $\boldsymbol{N}_1,\boldsymbol{N}_2$ and $\boldsymbol{N}_3$. are called "Lagrange triads." It is obvious that any unit vector $\boldsymbol{N}_i$ after rotation will changes to the unit vector $\boldsymbol{n}_i$, that is,

$$\boldsymbol{n}_i = \boldsymbol{R}\,\boldsymbol{N}_i \Rightarrow \boldsymbol{n}_i\boldsymbol{N}_i^T = \boldsymbol{R}\,\boldsymbol{N}_i\,\boldsymbol{N}_i^T$$

Since $\boldsymbol{N}_i \boldsymbol{N}_i^T = \boldsymbol{I}$ then we have:

$$\boldsymbol{R} = \boldsymbol{n}_i \boldsymbol{N}_i^T = \boldsymbol{n}_1 \boldsymbol{N}_1^T + \boldsymbol{n}_2 \boldsymbol{N}_2^T + \boldsymbol{n}_3 \boldsymbol{N}_3^T$$

The above expression can be written in terms of eigenvector matrices $\boldsymbol{q}(\boldsymbol{n})$ and $\boldsymbol{Q}(\boldsymbol{N})^T$:

$$\boldsymbol{R} = \boldsymbol{q}(\boldsymbol{n}) \boldsymbol{Q}(\boldsymbol{N})^T \tag{2.3.28}$$

If we substitute $\boldsymbol{R} = \boldsymbol{q}(\boldsymbol{n}) \boldsymbol{Q}(\boldsymbol{N})^T$ into $\boldsymbol{F} = \boldsymbol{R}\,\boldsymbol{U}$ then we have:

$$\boldsymbol{F} = \boldsymbol{q}(\boldsymbol{n}) \boldsymbol{Q}(\boldsymbol{N})^T\, \boldsymbol{U} \tag{2.3.29a}$$

Then substituting $\boldsymbol{U}$ from Eq. (2.3.20c) into Eq. (2.3.29a) provides this:

$$\boldsymbol{F} = \boldsymbol{q}(\boldsymbol{n}) \boldsymbol{Q}(\boldsymbol{N})^T\, \boldsymbol{Q}(\boldsymbol{N}) \operatorname{diag}(\lambda_1, \lambda_2, \lambda_3) \boldsymbol{Q}(\boldsymbol{N})^T$$

Since $\boldsymbol{Q}(\boldsymbol{N})^T\, \boldsymbol{Q}(\boldsymbol{N}) = \boldsymbol{I}$, the above expression will be simplified to:

$$\boldsymbol{F} = \boldsymbol{q}(\boldsymbol{n}) \operatorname{diag}(\lambda_1, \lambda_2, \lambda_3) \boldsymbol{Q}(\boldsymbol{N})^T \tag{2.3.29b}$$

The expansion of Eq. (2.3.29a) easily yields to:

$$\boldsymbol{F} = \lambda_1\, \boldsymbol{n}_1 \boldsymbol{N}_1^T + \lambda_2\, \boldsymbol{n}_2 \boldsymbol{N}_2^T + \lambda_3\, \boldsymbol{n}_3 \boldsymbol{N}_3^T \tag{2.3.29c}$$

The computational algorithm that, finds all the parameters from Eq. (2.2.6) or displacement gradient vector $\boldsymbol{D}$ can be summarized as follows:

a) From Eq. (2.2.6), which gives $\boldsymbol{D}$ we can find $\boldsymbol{F} = \boldsymbol{I} + \boldsymbol{D}$ and also $\boldsymbol{F}^T$ and finally right Cauchy Green tensor $\boldsymbol{F}^T \boldsymbol{F}$.
b) Then from Eq. (2.3.15) we can find λ_1^2, λ_2^2 and λ_3^2 and also $\boldsymbol{N}_1$, $\boldsymbol{N}_2$ and $\boldsymbol{N}_3$.

c) From equation (2.3.20b) and results of part b, the right stretch tensor $\boldsymbol{U}$ can be found.
d) Then by using $\boldsymbol{R}=\boldsymbol{F}^{-1}\boldsymbol{U}$ the rotation matrix can be found.
e) Since from part b the matrix $\boldsymbol{Q}(\boldsymbol{N})^T$ is known then from $\boldsymbol{R}=\boldsymbol{q}(\boldsymbol{n})\boldsymbol{Q}(\boldsymbol{N})^T$ the $\boldsymbol{q}(\boldsymbol{n})$ and subsequently the Euler triads $\boldsymbol{n}_1, \boldsymbol{n}_2$ and $\boldsymbol{n}_3$ could be found.
f) By results of part e, from Eq. (2.3.26) the left stretch tensor $\boldsymbol{V}$ can be found.
g) To check the accuracy the results of part a, part b and part e should be substituted into equation (2.3.29) and if it is satisfied the calculations are correct.

2.4 THE INVARIANTS OF THE STRAIN TENSOR AND THEIR TRANSFORMATION

In previous chapters, it was shown that the expressions for strain tensor in natural and orthogonal coordinate systems were different. In this section we want to study the transformation matrix that enables switching from one to another system. Moreover a two dimensional example will also be discussed.

The components of the strain tensor in natural coordinate system are designated by $\bar{\varepsilon}_{ij}$ such that the strain tensor $[\bar{\varepsilon}]$ can be shown by (see Eq. (2.1.8)):

$$[\varepsilon]=\bar{\varepsilon}_{ij}g^i g^j=\bar{\varepsilon}_{11}g^1 g^1+\bar{\varepsilon}_{12}g^1 g^2+\bar{\varepsilon}_{13}g^1 g^3+\bar{\varepsilon}_{21}g^2 g^1+\bar{\varepsilon}_{22}g^2 g^2 \\ +\bar{\varepsilon}_{23}g^2 g^3+\bar{\varepsilon}_{31}g^3 g^1+\bar{\varepsilon}_{32}g^3 g^2+\bar{\varepsilon}_{33}g^3 g^3 \tag{2.4.1a}$$

In section 2.2, we showed that we need to change $\vec{r}\Leftrightarrow X$ to be able to study the metric tensors in terms of orthogonal coordinate such the covariant base vector is (see Eq. (2.2.12)):

$$g_i=\frac{\partial \boldsymbol{X}}{\partial \alpha_i} \tag{2.4.1b}$$

The Jacobi matrix $\boldsymbol{J}$ can be defined by the elements containing covariant base vectors in Eq. (2.4.1b) into the following shape:

$$\boldsymbol{J} = \begin{bmatrix} \dfrac{\partial X_1}{\partial \alpha_1} & \dfrac{\partial X_2}{\partial \alpha_1} & \dfrac{\partial X_3}{\partial \alpha_1} \\ \dfrac{\partial X_1}{\partial \alpha_2} & \dfrac{\partial X_2}{\partial \alpha_2} & \dfrac{\partial X_3}{\partial \alpha_2} \\ \dfrac{\partial X_1}{\partial \alpha_3} & \dfrac{\partial X_2}{\partial \alpha_3} & \dfrac{\partial X_3}{\partial \alpha_3} \end{bmatrix} \tag{2.4.2}$$

Definition Eq. (2.4.2) says that $\boldsymbol{J}$ is a 3×3 matrix that its elements according to Eq. (2.4.1b) contains information about components of the base vectors $\boldsymbol{g}_1$, $\boldsymbol{g}_2$ and $\boldsymbol{g}_3$, in following form:

$$\boldsymbol{J} = \begin{bmatrix} g_1(1) & g_1(2) & g_1(3) \\ g_2(1) & g_2(2) & g_2(3) \\ g_3(1) & g_3(2) & g_3(3) \end{bmatrix} \tag{2.4.3}$$

Each base vector has 3 components in orthogonal system, and the 3 base vectors have 9 component in total, and these components are also the elements of the Jacobi matrix. According to Eq. (2.2.12a) these components can be written into this form:

$$g_r = g_r(1)\vec{i}_1 + g_r(2)\vec{i}_2 + g_r(3)\vec{i}_3 = g_r(i)\vec{i}_i \qquad i = 1,2,3 \tag{2.4.4}$$

Now we form the scalar multiplication of the both sides into the unit vector $\vec{i}_s$, then we have:

$$g_r \cdot \vec{i}_s = g_r(1)\vec{i}_1 + g_r(2)\vec{i}_2 + g_r(3)\vec{i}_3 = g_r(i)\vec{i}_i \cdot \vec{i}_s$$

In the above expression $\vec{i}_i \cdot \vec{i}_s = 1 \quad i = s$ and $\vec{i}_i \cdot \vec{i}_s = 0 \quad i \neq s$, therefore the results will change to $g_r(s)$ which is the elements of Jacobi matrix J_{rs}, all statements above can be summarized by these formulas:

$$\boldsymbol{J} = [J_{rs}] = [g_r \cdot \vec{i}_s] \quad J_{rs} = g_r(s) \quad r,s = 1,2,3 \tag{2.4.5}$$

Now we use different symbol and show the components of the strain tensor in orthogonal coordinate by $\hat{\varepsilon}_{rs}$, such that a strain tensor can be shown in two ways like this:

$$[\varepsilon] = \bar{\varepsilon}_{rs} g^r g^s = \hat{\varepsilon}_{rs} i_r i_s \tag{2.4.6}$$

We can change Eq. (2.4.6) which is written in tensor form with a new equation in vector form and for doing this we form the scalar product of both sides with g_k such that:

$$\bar{\varepsilon}_{rs} \underbrace{(g_k \cdot g^r)}_{\delta_k^r} g^s = \hat{\varepsilon}_{rs} \underbrace{(g_k \cdot i_r)}_{J_{kr}} i_s \Rightarrow \delta_k^r \bar{\varepsilon}_{rs} g^s = J_{kr} \hat{\varepsilon}_{rs} i_s \tag{2.4.7a}$$

Equation (2.4.7a) is written in vector form and if we form the scalar product of both sides with g_k once more, it will change to scalar form, that is,

$$\delta_k^r \bar{\varepsilon}_{rs} \underbrace{(g_k \cdot g^s)}_{\delta_k^s} = J_{kr} \hat{\varepsilon}_{rs} \underbrace{(g_k \cdot i_s)}_{J_{ks}} \Rightarrow \delta_k^r \bar{\varepsilon}_{rs} \delta_k^s = J_{kr} \hat{\varepsilon}_{rs} J_{ks} \tag{2.4.7b}$$

The indicial Eq. (2.4.7b) is scalar and since δ_k^r and δ_k^s are either 1 or 0, we call the matrix containing elements $[\bar{\varepsilon}_{rs}]$ by $\bar{\varepsilon}$ and the one containing elements $[\hat{\varepsilon}_{rs}]$ by $\hat{\boldsymbol{\varepsilon}}$. Then Eq. (2.4.7b) can be written into one of these forms:

$$[\bar{\varepsilon}_{rs}] = [J_{kr}][\hat{\varepsilon}_{rs}][J_{ks}] \tag{2.4.8}$$

$$\bar{\boldsymbol{\varepsilon}} = \boldsymbol{J} \hat{\boldsymbol{\varepsilon}} \boldsymbol{J}^T \tag{2.4.9}$$

Now we set an example to understand the difference between natural coordinate and orthogonal coordinate system. For sake of simplicity we choose a two dimensional case and this example is also related to the analysis of isoperimetric elements in finite element method, whereby for skewed elements the natural coordinate is chosen as indicated in Fig. 2.4, by natural axis α^I and α^{II} which represents a parallelepiped with dimensions $2a$ and $2b$

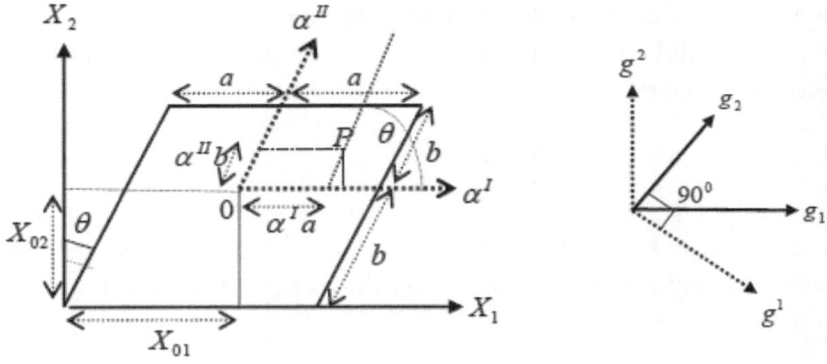

FIGURE 2.4 The natural coordinates α^I and α^{II}.

According to Fig. 2.4, any point P in the body can be expressed by the orthogonal coordinates like:

$$\vec{X} = X_1\vec{i}_1 + X_2\vec{i}_2 \tag{2.4.10a}$$

$$X_1 = a\alpha^I + b\alpha^{II}\cos\theta + X_0 \qquad X_2 = b\alpha^{II}\sin\theta + Y_0 \tag{2.4.10b}$$

The covariant base vectors will be:

$$g_1 = \frac{\partial X}{\partial \alpha^I} = \frac{\partial X_1}{\partial \alpha^I}\vec{i}_1 + \frac{\partial X_2}{\partial \alpha^I}\vec{i}_2$$

$$\frac{\partial X_1}{\partial \alpha^I} = a \qquad \frac{\partial X_2}{\partial \alpha^I} = 0 \qquad g_1 = a\vec{i}_1 \tag{2.4.11a}$$

$$g_2 = \frac{\partial X}{\partial \alpha^{II}} = \frac{\partial X_1}{\partial \alpha^{II}}\vec{i}_1 + \frac{\partial X_2}{\partial \alpha^{II}}\vec{i}_2$$

$$\frac{\partial X_1}{\partial \alpha^{II}} = b\cos\theta \quad \frac{\partial X_2}{\partial \alpha^{II}} = b\sin\theta \quad g_2 = b\cos\theta\,\vec{i}_1 + b\sin\theta\,\vec{i}_2 \tag{2.4.11b}$$

Therefore, we can find the components of the metric tensor like:

$$g_{11} = g_1 \cdot g_1 = a^2 \qquad g_{22} = g_2 \cdot g_2 = b^2 \cos^2\theta + b^2 \sin^2\theta = b^2$$

$$g_{12} = g_1 \cdot g_2 = ab\cos\theta$$

Then the covariant geometric tensor is:

$$\left[g_{ij}\right] = \begin{bmatrix} a^2 & ab\cos\theta \\ ab\cos\theta & b^2 \end{bmatrix} \qquad |g| = \det\left[g_{ij}\right] = a^2b^2\sin^2\theta$$

Also the contra-variant geometric tensor can be found by using Eq. (1.2.7) so that:

$$\left[g^{ij}\right] = \left[g_{ij}\right]^{-1} = \frac{1}{a^2b^2\sin^2\theta}\begin{bmatrix} b^2 & -ab\cos\theta \\ -ab\cos\theta & a^2 \end{bmatrix}$$

Moreover, the contra-variant base vectors can be found from Eq. (1.2.5) like this:

$$g^1 = g^{11}g_1 + g^{12}g_2 = \frac{1}{a^2\sin^2\theta}\begin{bmatrix} a \\ 0 \end{bmatrix} + \frac{-\cos\theta}{ab\sin^2\theta}\begin{bmatrix} b\cos\theta \\ b\sin\theta \end{bmatrix}$$

Exercise 2.4.1: *Simplify* g^1 in the above:

Solution:

$$g^1 = \frac{1}{a^2\sin^2\theta}\begin{bmatrix} a \\ 0 \end{bmatrix} + \frac{-\cos\theta}{ab\sin^2\theta}\begin{bmatrix} b\cos\theta \\ b\sin\theta \end{bmatrix} = \frac{1}{a^2b\sin^2\theta}\begin{bmatrix} ab - ab\cos^2\theta \\ ab\sin\theta \end{bmatrix}$$ and

this can be simplified to: $g^1 = \frac{1}{a}\begin{bmatrix} 1 \\ \text{cotg}\theta \end{bmatrix}$

$$g^1 = g^{21}g_1 + g^{22}g_2 = \frac{-\cos\theta}{ab\sin^2\theta}\begin{bmatrix} a \\ 0 \end{bmatrix} + \frac{1}{b^2\sin^2\theta}\begin{bmatrix} b\cos\theta \\ b\sin\theta \end{bmatrix}$$

And can be simplified

$$g^1 = \frac{-\cos\theta}{ab\sin^2\theta}\begin{bmatrix} a \\ 0 \end{bmatrix} + \frac{1}{b^2\sin^2\theta}\begin{bmatrix} b\cos\theta \\ b\sin\theta \end{bmatrix} = \frac{1}{ab^2\sin^2\theta}\begin{bmatrix} 0 \\ ab\sin\theta \end{bmatrix}$$ and further simplification gives: $g^1 = \frac{1}{b\sin\theta}\begin{bmatrix} 0 \\ 1 \end{bmatrix}$

If we want to find the metric tensor in deformed body we need to change R to x so that the covariant base vectors will be:

$$G_i = \frac{\partial \boldsymbol{x}}{\partial \alpha^i} = \frac{\partial(\boldsymbol{X} + \boldsymbol{d})}{\partial \alpha^i} = g_i + \frac{\partial \boldsymbol{d}}{\partial \alpha^i}$$

Herein for simplicity we show displacement vector with $\boldsymbol{d} = \begin{bmatrix} u \\ v \end{bmatrix}$ and the natural coordinate $\alpha^I = \zeta$ and $\alpha^{II} = \eta$ such that $\frac{\partial \boldsymbol{d}}{\partial \alpha^I} = \begin{bmatrix} u_\zeta \\ v_\zeta \end{bmatrix}$ and $\frac{\partial \boldsymbol{d}}{\partial \alpha^{II}} = \begin{bmatrix} u_\eta \\ v_\eta \end{bmatrix}$ in which:

$$u_\zeta = \frac{\partial u}{\partial \alpha^I} \qquad v_\zeta = \frac{\partial v}{\partial \alpha^I} \qquad u_\eta = \frac{\partial u}{\partial \alpha^{II}} \qquad v_\eta = \frac{\partial v}{\partial \alpha^{II}}$$

Therefore, we have:

$$G_1 = g_1 + \begin{bmatrix} u_\zeta \\ v_\zeta \end{bmatrix} = \begin{bmatrix} a + u_\zeta \\ v_\zeta \end{bmatrix} \qquad G_2 = g_2 + \begin{bmatrix} u_\eta \\ v_\eta \end{bmatrix} = \begin{bmatrix} b\cos\theta + u_\eta \\ b\sin\theta + v_\eta \end{bmatrix}$$

The elements of the metric tensor (deformed) then are:

$$G_{11} = G_1 \cdot G_1 = \left(a + u_\zeta\right)^2 + v_\zeta^2$$

$$G_{22} = G_2 \cdot G_2 = \left(b\cos\theta + u_\eta\right)^2 + \left(b\sin\theta + v_\eta\right)^2$$

$$G_{12} = G_1 \cdot G_2 = \left(a + u_\zeta\right)\left(b\cos\theta + u_\eta\right) + \left(b\sin\theta + v_\eta\right)v_\zeta$$

The components of the strain tensor are:

$$\bar{\varepsilon}_{11} = \frac{1}{2}(G_{11} - g_{11}) = \frac{1}{2}\left(\left(a + u_\zeta\right)^2 + v_\zeta^2 - a^2\right) = \frac{1}{2}\left(u_\zeta^2 + v_\zeta^2\right) + au_\zeta$$

$$\bar{\varepsilon}_{22} = \frac{1}{2}(G_{22} - g_{22}) = \frac{1}{2}\begin{pmatrix} (b\cos\theta + u_\eta)^2 + \\ (b\sin\theta + v_\eta)^2 - b^2 \end{pmatrix} = \frac{1}{2}(u_\eta^2 + v_\eta^2) + b\begin{pmatrix} u_\eta \cos\theta + \\ v_\eta \sin\theta \end{pmatrix}$$

$$\bar{\varepsilon}_{12} = \frac{1}{2}(G_{12} - g_{12}) = \frac{1}{2}\left((a + u_\varsigma)(b\cos\theta + u_\eta) + (b\sin\theta + v_\eta)v_\varsigma - ab\cos\theta\right)$$

Now we are in the situation that can find elements of the Jacobin matrix by using Eq. (2.4.5) such that:

$$J_{11} = [g_1 \cdot i_1] = a i_1 \cdot i_1 = a \quad J_{21} = [g_2 \cdot i_1] = (b\cos\theta\, \vec{i}_1 + b\sin\theta\, \vec{i}_2) \cdot i_1 = b\cos\theta$$

$$J_{12} = [g_1 \cdot i_2] = a i_1 \cdot i_2 = 0 \quad J_{22} = [g_2 \cdot i_2] = (b\cos\theta\, \vec{i}_1 + b\sin\theta\, \vec{i}_2) \cdot i_2 = b\sin\theta$$

Now that Jacobi matrix is $\boldsymbol{J} = \begin{bmatrix} a & 0 \\ b\cos\theta & b\sin\theta \end{bmatrix}$ its inverse is $\boldsymbol{J}^{-1} = \frac{1}{ab\sin\theta}\begin{bmatrix} b\sin\theta & 0 \\ -b\cos\theta & a \end{bmatrix}$. The equation (2.4.9) can be changed first by pre-multiplying to $\boldsymbol{J}^{-1}$ and then by post-multiplying to $\boldsymbol{J}^{-T}$ which results a new expression like this:

$$\hat{\boldsymbol{\varepsilon}} = \boldsymbol{J}^{-1}\bar{\boldsymbol{\varepsilon}}\,\boldsymbol{J}^{-T} \tag{2.4.10}$$

The elements of Green strain tensor in natural coordinate system $\bar{\varepsilon}_{11}$, $\bar{\varepsilon}_{22}$ and $\bar{\varepsilon}_{12}$ are already computed. These elements in orthogonal coordinate system are $\hat{\varepsilon}_{11}$, $\hat{\varepsilon}_{22}$ and $\hat{\varepsilon}_{12}$ which can be found by substituting $\boldsymbol{J}^{-1}$ in Eq. (2.4.10) such that:

$$\begin{bmatrix} \hat{\varepsilon}_{11} & \hat{\varepsilon}_{12} \\ \hat{\varepsilon}_{21} & \hat{\varepsilon}_{22} \end{bmatrix} = \frac{1}{a^2 b^2 \sin^2\theta}\begin{bmatrix} b\sin\theta & 0 \\ -b\cos\theta & a \end{bmatrix}\begin{bmatrix} \bar{\varepsilon}_{11} & \bar{\varepsilon}_{12} \\ \bar{\varepsilon}_{21} & \bar{\varepsilon}_{22} \end{bmatrix}\begin{bmatrix} b\sin\theta & -b\cos\theta \\ 0 & a \end{bmatrix}$$

After multiplication and simplification and rearrangement we have:

$$\hat{\varepsilon}_{11} = \frac{\bar{\varepsilon}_{11}}{a^2} \qquad \hat{\varepsilon}_{12} = \hat{\varepsilon}_{21} = -\bar{\varepsilon}_{11}\frac{\cot\theta}{a^2} + \frac{\bar{\varepsilon}_{22}}{ab\sin\theta}$$

$$\hat{\varepsilon}_{22} = \bar{\varepsilon}_{11}\frac{\cot^2\theta}{a^2} - 2\bar{\varepsilon}_{12}\frac{\cos\theta}{a\,b\sin^2\theta} + \frac{\bar{\varepsilon}_{22}}{b^2\sin^2\theta}$$

In orthogonal systems, under transformations there are invariants that do not change and for the stress and also strain tensor we studied this as a eigenvalue problem for small strains that can be found from the following determinant:

$$\det\left(\varepsilon - \varepsilon_p \boldsymbol{I}\right) = 0 \tag{2.4.11}$$

By expansion of the determinant in Eq. (2.3.11) the principal strains ε_p can be found from:

$$\varepsilon_p^3 + I_1\varepsilon_p^2 + I_2\,\varepsilon_p + I_3 = 0 \tag{2.4.12}$$

The strain invariant I_1, I_2 and I_3 are:

$$I_1 = \varepsilon_{11} + \varepsilon_{22} + \varepsilon_{33} = \mathrm{tr}\left(\varepsilon\right) \quad I_3 = \det\left(\varepsilon\right) \tag{2.4.13}$$

$$I_2 = \varepsilon_{11}\,\varepsilon_{22} + \varepsilon_{22}\,\varepsilon_{33} + \varepsilon_{33}\,\varepsilon_{11} - \varepsilon_{12}^2 - \varepsilon_{13}^2 - \varepsilon_{23}^2 = \frac{1}{2}\left(\mathrm{tr}\left(\varepsilon\right)^2 - \mathrm{tr}\left(\varepsilon\,\varepsilon\right)\right)$$

Equations (2.4.11), (2.4.12) and (2.4.13) was also studied in elasticity and it is also common practice to use them in plasticity. However, in hyper-elasticity we use the stretches λ_1, λ_2 and λ_3 that are the parameters always greater the one.

Moreover, in previous section we introduced right Cauchy-Green stretch tensor by $\boldsymbol{b} = \boldsymbol{F}^T\boldsymbol{F}$ and also left stretch tensor $\boldsymbol{c} = \boldsymbol{F}\,\boldsymbol{F}^T$ and from them we can find out the squares of principal stretches λ_1^2, λ_2^2 and λ_3^2 by using either Eqs. (2.3.15) or (2.3.24) that gives the stretches as the eigen-values. Then the stretch invariants can be defined similar to Eq. (2.4.13) into this form:

$$I_1 = \lambda_1^2 + \lambda_2^2 + \lambda_3^2 = c_{11} + c_{22} + c_{33} = \mathrm{tr}\left(\boldsymbol{c}\right) = \boldsymbol{I} : \boldsymbol{c} = \mathrm{tr}\left(\boldsymbol{b}\right) = \boldsymbol{I} : \tag{2.4.14}$$

Question 2.4.1: Explain what is $\boldsymbol{I}:\boldsymbol{b}$ and $\boldsymbol{I}:\boldsymbol{c}$?

Answer: It is based on dot (inner) product of two matrices in which the result is a scalar value. The Frobenius inner product of two matrices is $\boldsymbol{A}:\boldsymbol{B}=\sum_{i,j=1}^{n} A_{ij}B_{ij}$ and using this rule:

$$\boldsymbol{I}:\boldsymbol{b}=\begin{bmatrix}1&0&0\\0&1&0\\0&0&1\end{bmatrix}:\begin{bmatrix}b_{11}&b_{12}&b_{13}\\b_{21}&b_{22}&b_{23}\\b_{31}&b_{32}&b_{33}\end{bmatrix}=1\times b_{11}+0\times b_{12}+0\times b_{13}+0\times b_{21}+1\times b_{22}$$
$$+0\times b_{23}+0\times b_{31}+0\times b_{32}+1\times b_{33}$$
$$=b_{11}+b_{22}+b_{33}=\operatorname{tr}(\boldsymbol{b})$$

According to the Eqs. (2.2.16) and (2.2.17) we can write:

$$\boldsymbol{F}=[G_i][g^i]^T \quad \boldsymbol{F}^T=[g^j][G_j]^T$$

Also by definition in the chapter one, we had $[g^i]^T[g^j]=g^{ij}$, then by substituting these three expressions into $\boldsymbol{b}=\boldsymbol{F}^T\boldsymbol{F}$, then we have:

$$\boldsymbol{b}=g^{ij}[G_i][G_j]^T \quad i,j=1,2,3 \tag{2.4.15}$$

The above expression in Eq. (2.4.15) consists of 9 matrices, therefore we need to add the traces of 9 matrices together to find $\boldsymbol{I}:\boldsymbol{b}$, the following exercise shows the route:

Exercise 2.4.2: *Find* $\boldsymbol{I}:\left(g^{12}[G_1][G_2]^T\right)$

Solution: $$\boldsymbol{I}:\left(g^{12}[G_1][G_2]^T\right)=g^{12}\boldsymbol{I}:\begin{bmatrix}G_1(1)\\G_1(2)\\G_1(3)\end{bmatrix}\begin{bmatrix}G_2(1)&G_2(2)&G_2(3)\end{bmatrix}$$

and this can be rewritten into

$$\boldsymbol{I}:\left(g^{12}[G_1][G_2]^T\right)=g^{12}\begin{bmatrix}1&0&0\\0&1&0\\0&0&1\end{bmatrix}:\begin{bmatrix}G_1(1)G_2(1)&G_1(1)G_2(2)&G_1(1)G_2(3)\\G_1(2)G_2(1)&G_1(2)G_2(2)&G_1(2)G_2(3)\\G_1(3)G_2(1)&G_1(3)G_2(2)&G_1(3)G_2(3)\end{bmatrix}$$

and this yields to $\boldsymbol{I}:\left(g^{12}[G_1][G_2]^T\right)=g^{12}\left(G_1(1)G_2(1)+G_1(1)G_2(1)+G_1(1)G_2(1)\right)=g^{12}\left(G_1\cdot G_2\right)$ and the final form will be $\boldsymbol{I}:\left(g^{12}[G_1][G_2]^T\right)=g^{12}G_{12}$

Therefore $\boldsymbol{I}:\boldsymbol{b}$ contains 9 terms like above, that is,

$$\begin{aligned}\boldsymbol{I}:\boldsymbol{b}&=g^{11}G_{11}+g^{12}G_{12}+g^{13}G_{13}+g^{21}G_{21}+g^{22}G_{22}+\\&+g^{23}G_{23}+g^{31}G_{31}+g^{32}G_{32}+g^{33}G_{33}\end{aligned}$$

Finally the first invariant of the stretch tensor which is defined by Eq. (2.4.12) can be expressed in indicial form like this:

$$I_1=\boldsymbol{I}:\boldsymbol{b}=g^{rs}G_{rs}\quad r,s=1,2,3 \tag{2.4.16}$$

The 2nd invariant of the stretch tensor can be expressed by:

$$I_2=\lambda_1^2\lambda_2^2+\lambda_3^2\lambda_2^2+\lambda_1^2\lambda_3^2=b_{11}b_{22}+b_{11}b_{33}+b_{33}b_{22}-b_{12}b_{21}-b_{13}b_{31}-b_{23}b_{32} \tag{2.4.17a}$$

Exercise 2.4.3: *Simplify* Eq. (2.4.17a)

Solution: The first diagonal element of the matrix $(\boldsymbol{b}\,\boldsymbol{b})$ is $b_{11}^2+b_{12}b_{21}+b_{13}b_{31}$, the 2nd diagonal element is $b_{22}^2+b_{12}b_{21}+b_{23}b_{32}$ and the 3rd element is $b_{33}^2+b_{13}b_{31}+b_{23}b_{32}$, therefore the trace of the matrix $\mathrm{tr}(\boldsymbol{b}\,\boldsymbol{b})=b_{11}^2+b_{22}^2+b_{33}^2+2b_{12}b_{21}+2b_{13}b_{31}+2b_{32}b_{23}$. However, from Eq. (2.4.14) we have $I_1^2=(b_{11}+b_{22}+b_{33})^2$ and if we substitute all into Eq. (2.4.17a) then we have:

$$I_2=\frac{1}{2}\left(I_1^2-\mathrm{tr}(\boldsymbol{b}\,\boldsymbol{b})\right) \tag{2.4.17b}$$

By substituting Eq. (1.2.4), that is, $[G_i]=G_{ir}\left[G^r\right]$ into Eq. (2.4.15) it will change to:

$$\boldsymbol{b}=g^{ij}G_{ir}\left[G^r\right]\left[G_j\right]^T\quad i,j,r=1,2,3 \tag{2.4.18}$$

Exercise 2.4.4: *By using* Eq. (2.4.18) *find an indicial expression for* $\mathrm{tr}(\boldsymbol{b}\boldsymbol{b})$:

Solution: Once by using Eq. (2.4.18) we write $\boldsymbol{b} = g^{rm} G_{mn}\left[G^r\right]\left[G_n\right]^T$ the we change indies and write $\boldsymbol{b} = g^{ps} G_{sq}\left[G^p\right]\left[G_q\right]^T$ then we can form $(\boldsymbol{b}\boldsymbol{b})$ as follows:

$$(\boldsymbol{b}\boldsymbol{b}) = g^{rm} G_{mn} g^{ps} G_{sq}\left[G^p\right]\underbrace{\left[G_q\right]^T\left[G^r\right]}_{\delta_{qr}}\left[G_n\right]^T$$

However, according to Eq. (1.1.6) we have $\left[G_q\right]^T\left[G^r\right] = \delta_{qr}$ and the above expression will be:

$$(\boldsymbol{b}\boldsymbol{b}) = g^{rm} G_{mn} g^{ps} G_{sq} \delta_{qr}\left[G^p\right]\left[G_n\right]^T$$

Considering that $G_{sq}\delta_{qr} = G_{sr}$ the above expression will change to;

$$(\boldsymbol{b}\boldsymbol{b}) = g^{rm} G_{mn} g^{ps} G_{sr}\left[G^p\right]\left[G_n\right]^T$$

By repeating exercise Eq. (2.4.2) for the above expression we can find that:

$$\boldsymbol{I}:(\boldsymbol{b}\boldsymbol{b}) = \boldsymbol{I}:\left(g^{rm} G_{mn} g^{ps} G_{sr}\left[G^p\right]\left[G_n\right]^T\right) = g^{rm} G_{mn} g^{ps} G_{sr}\left(G^p \cdot G_n\right)$$

Again according to (1.1.6) $G^p \cdot G_n = \delta_{pn}$ then we have:

$$\boldsymbol{I}:(\boldsymbol{b}\boldsymbol{b}) = g^{rm} G_{mn} g^{ps} G_{sr} \delta_{pn}$$

Considering that $g^{ps}\delta_{pn} = g^{ns}$ the above expression will change to:

$$\mathrm{tr}(\boldsymbol{b}\boldsymbol{b}) = \boldsymbol{I}:(\boldsymbol{b}\boldsymbol{b}) = g^{rm} g^{ns} G_{mn} G_{sr} \quad m,n,s,r = 1,2,3$$

Substituting the above expression into Eq. (2.4.17b) provides 2nd invariant of the stretch tensor as:

$$I_2 = \frac{1}{2}\left(I_1^2 - g^{rm} g^{ns} G_{mn} G_{sr}\right) \quad m,n,s,r = 1,2,3 \tag{2.4.17c}$$

Substituting Eq. (2.4.16) into Eq. (2.4.17c) give another form for I_2 which is:

$$I_2 = \frac{1}{2}\left(\left(g^{rs} G_{rs}\right)^2 - g^{rm} g^{ns} G_{mn} G_{sr}\right) \quad m,n,s,r = 1,2,3 \tag{2.4.17d}$$

The 3rd invariant of the stretch tensor can be defined by:

$$I_3 = \lambda_1^2 \lambda_2^2 \lambda_3^2 = \det(\boldsymbol{b}) = \det(\boldsymbol{c}) = J^2 \tag{2.4.19a}$$

The parameter $J = \sqrt{I_3}$ in Eq. (2.4.19a) is called Jacobian and is related to volume change so that for incompressible material $J = 1$, to find out what is $\det(\boldsymbol{b})$? Rewrite Eq. (2.4.18) into $\boldsymbol{b} = g^{rm} G_{mn} \left[G^r\right]\left[G_n\right]^T$ and then we post multiply both sides into $\left[G^s\right]$ and we have:

$$\boldsymbol{b}\left[G^s\right] = g^{rm} G_{mn} \left[G^r\right] \underbrace{\left[G_n\right]^T \left[G^s\right]}_{\delta_{ns}}$$

Since $\left[G_n\right]^T \left[G^s\right] = \delta_{ns}$, then the above expression changes to:

$$\boldsymbol{b}\left[G^s\right] = g^{rm} G_{mn} \left[G^r\right] \delta_{ns}$$

Also we have $G_{mn}\delta_{ns} = G_{ms}$ and the above expression can be simplified to:

$$\boldsymbol{b}\left[G^s\right] = g^{rm} G_{ms} \left[G^r\right]$$

Further multiplication of the above equation into $\left[G_s\right]^T$ provides:

$$\boldsymbol{b}\left[G^s\right]\left[G_s\right]^T = g^{rm} G_{ms} \left[G^r\right]\left[G_s\right]^T \tag{2.4.19b}$$

The left side of Eq. (2.4.19b) is only 3 matrices but the right side is 9 matrices and the following exercise clarifies what they are?

Exercise 2.4.5: *Find* $\boldsymbol{b}\left[G^s\right]\left[G_s\right]^T$ *in non-orthogonal base of the deformed body*

Solution: $$\boldsymbol{b}\left[G^s\right]\left[G_s\right]^T=\begin{bmatrix} b_{11} & b_{12} & b_{13} \\ b_{21} & b_{22} & b_{23} \\ b_{31} & b_{32} & b_{33} \end{bmatrix}\left[\begin{bmatrix}1\\0\\0\end{bmatrix}\begin{bmatrix}1 & 0 & 0\end{bmatrix}+\begin{bmatrix}0\\1\\0\end{bmatrix}\begin{bmatrix}0 & 1 & 0\end{bmatrix}+\begin{bmatrix}0\\0\\1\end{bmatrix}\begin{bmatrix}0 & 0 & 1\end{bmatrix}\right]$$

and results $$\boldsymbol{b}\left[G^s\right]\left[G_s\right]^T=\begin{bmatrix} b_{11} & b_{12} & b_{13} \\ b_{21} & b_{22} & b_{23} \\ b_{31} & b_{32} & b_{33} \end{bmatrix}\left[\begin{bmatrix}1 & 0 & 0\\0 & 0 & 0\\0 & 0 & 0\end{bmatrix}+\begin{bmatrix}0 & 0 & 0\\0 & 1 & 0\\0 & 0 & 0\end{bmatrix}+\begin{bmatrix}0 & 0 & 0\\0 & 0 & 0\\0 & 0 & 1\end{bmatrix}\right]$$ and

after adding inside the bracket we have:

$$\boldsymbol{b}\left[G^s\right]\left[G_s\right]^T=\begin{bmatrix} b_{11} & b_{12} & b_{13} \\ b_{21} & b_{22} & b_{23} \\ b_{31} & b_{32} & b_{33} \end{bmatrix}\begin{bmatrix}1 & 0 & 0\\0 & 1 & 0\\0 & 0 & 1\end{bmatrix}=\boldsymbol{b} \qquad (2.4.19c)$$

Exercise 2.4.6: *Find the matrix for* $g^{2m}G_{m3}\left[G^2\right]\left[G_3\right]^T$

Solution: $$g^{2m}G_{m3}\left[G^2\right]\left[G_3\right]^T=g^{2m}G_{m3}\begin{bmatrix}0\\1\\0\end{bmatrix}\begin{bmatrix}0 & 0 & 1\end{bmatrix}=g^{2m}G_{m3}\begin{bmatrix}0 & 0 & 0\\0 & 0 & 1\\0 & 0 & 0\end{bmatrix}$$

$$=\begin{bmatrix}0 & 0 & 0\\0 & 0 & g^{2m}G_{m3}\\0 & 0 & 0\end{bmatrix}$$

This exercise can be repeated 9 times and it can be shown that:

$$g^{rm}G_{ms}\left[G^{r}\right]\left[G_{s}\right]^{T}=\begin{bmatrix} g^{1m}G_{m1} & g^{1m}G_{m2} & g^{1m}G_{m3} \\ g^{2m}G_{m1} & g^{2m}G_{m2} & g^{2m}G_{m3} \\ g^{3m}G_{m1} & g^{3m}G_{m2} & g^{3m}G_{m3} \end{bmatrix} \quad r,s=1,2,3 \tag{2.4.19d}$$

Substituting Eqs. (2.4.19c) and (2.4.19d) into Eq. (2.4.19b) results this expression:

$$\boldsymbol{b}=\begin{bmatrix} g^{1m}G_{m1} & g^{1m}G_{m2} & g^{1m}G_{m3} \\ g^{2m}G_{m1} & g^{2m}G_{m2} & g^{2m}G_{m3} \\ g^{3m}G_{m1} & g^{3m}G_{m2} & g^{3m}G_{m3} \end{bmatrix} \tag{2.4.20}$$

The matrix in Eq. (2.4.20) contains the elements of product of two matrices $\bar{\boldsymbol{g}}=\left[g^{rm}\right]$ and $\boldsymbol{G}=\left[G_{ms}\right]$ such that:

$$\boldsymbol{b}=\left[g^{rm}\right]\left[G_{ms}\right]=\bar{\boldsymbol{g}}\boldsymbol{G} \tag{2.4.21}$$

By pre-multiplying both sides of Eq. (2.4.21) into $\left[g_{rm}\right]$ we have:

$$\boldsymbol{g}\,\boldsymbol{b}=\left[g_{rm}\right]\left[g^{rm}\right]\left[G_{ms}\right]=\boldsymbol{g}\,\bar{\boldsymbol{g}}\,\boldsymbol{G}$$

According to Eq. (1.2.7) we have $\left[g_{rm}\right]\left[g^{rm}\right]=\boldsymbol{g}\,\bar{\boldsymbol{g}}=\boldsymbol{I}$, therefore the above expression will change to:

$$\boldsymbol{g}\,\boldsymbol{b}=\left[G_{ms}\right]=\boldsymbol{G} \tag{2.4.22}$$

Taking the determinant of the matrix expression in Eq. (2.4.22) results:

$$\det(\boldsymbol{g})\det(\boldsymbol{b})=\det\left[G_{ms}\right]=\det(\boldsymbol{G})$$

Comparing the above expression with Eq. (2.4.19a) results:

$$I_3=\det(\boldsymbol{b})=G/g \tag{2.4.23}$$

In Eq. (2.4.23), $G = \det(\boldsymbol{G})$ and $g = \det(\boldsymbol{g})$ when the material is incompressible we have $G = g$ and therefore $\frac{G}{g} = 1$, therefore I_3 can be accounted as a compressibility criteria. This can be easily demonstrated if we look at Eq. (1.3.8) we have:

$$dv_0 = \sqrt{g}\ dx^1\, dx^2 dx^3$$

$$dv = \sqrt{G}\ dx^1\, dx^2 dx^3$$

And by a simple division we have:

$$J = \frac{dv}{dv_0} = \sqrt{\frac{G}{J}} \Rightarrow \sqrt{I_3} = \sqrt{\frac{G}{J}} \Rightarrow I_3 = \frac{G}{J}$$

KEYWORDS

- **contra-variant geometric tensor**
- **finite element method**
- **Green strain tensor**
- **Jacobi matrix**
- **left Cauchy Green tensor**
- **orthogonal systems**

CHAPTER 3

GENERAL THEORY FOR STRESS IN SOLIDS

CONTENTS

3.1 STRESS TENSOR AND ITS FORMULA

According to Fig. 3.1, in each point P in the solid, three curves intersect each other. The first curve is θ_3 designated by PP_3 and in the deformed solid the covariant-based vector G_3 is tangent to PP_3. Similarly 2nd curve is θ_2 designated by PP_2 and G_2 is tangent to PP_2, also 3rd curve is θ_1 designated by PP_1 and G_1 is tangent to PP_1. It should be reminded that in this section

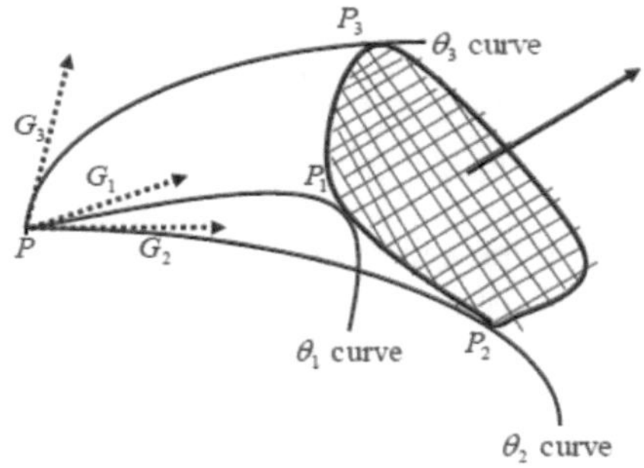

FIGURE 3.1 Stress in a point P.

we discuss the general definition of the stress in non-orthogonal coordinate system and the triangle $P_1P_2P_3$ represents a area in the deformed solid.

We should emphasize that an accurate stress on an area $P_1P_2P_3$ can be defined only on a deformed solid. This route will be followed first and later in the section we define unrealistic stresses as well. The accurate stress, which is defined in the deformed solid in non-orthogonal coordinate system is called Green stress tensor.

The curves PP_1, PP_2 and PP_3 are the differentials of the length $d\alpha_1$, $d\alpha_2$ and $d\alpha_3$ which are defined over the curves θ_1, θ_2 and θ_3. In the prism $PP_1P_2P_3$ there are three side surfaces each have the surfaces on the deformed body as follows:

$$\overrightarrow{PP_1P_2} = \frac{G^3 dS_3}{2\sqrt{G^{33}}} \quad \overrightarrow{PP_1P_3} = \frac{G^2 dS_2}{2\sqrt{G^{22}}} \quad \overrightarrow{PP_2P_3} = \frac{G^1 dS_1}{2\sqrt{G^{11}}} \tag{3.1.1}$$

The part equations in Eq. (3.1.1) are based on the formula for area in chapter one given by Eq. (1.3.3). For example, triangle PP_1P_2 is formed by the curves θ_1 and θ_2 it is called the surface $\theta_3 = \text{constant}$ and is perpendicular to the covariant-based vectors G_1 and G_2 therefore it will be in the direction G^3. The coefficient 2, in the denominator is due to triangle area. The vector $\overrightarrow{P_1P_2P_3}$ that represent the are of the triangle in Fig. 3.1, can be written by:

$$\overrightarrow{P_1P_2P_3} = \overrightarrow{PP_1P_2} + \overrightarrow{PP_1P_3} + \overrightarrow{PP_3P_2} \tag{3.1.2}$$

If the unit normal r to the area $P_1P_2P_3$ by value $\frac{1}{2}dS$ is N, then $\overrightarrow{P_1P_2P_3} = \frac{1}{2}N\,dS$, by substituting this and Eq. (3.1.2) into Eq. (3.1.1) we have:

$$N\,dS = \sum_{i=1}^{3} \frac{G^i dS_i}{\sqrt{G^{ii}}} \tag{3.1.3}$$

From Eq. (3.1.3) it is obvious that G^1, G^2 and G^3 are involved in the formulas. The covariant components of $\vec{N}$ (unit normal) over the contravariant base vectors G^1, G^2 and G^3 is:

$$N = N_1G^1 + N_2G^2 + N_3G^3 \tag{3.1.4}$$

Substituting Eq. (3.1.4) yields to:

$$\left(N_1G^1+N_2G^2+N_3G^3\right)dS=\frac{G^1dS_1}{\sqrt{G^{11}}}+\frac{G^2dS_2}{\sqrt{G^{22}}}+\frac{G^3dS_3}{\sqrt{G^{33}}}$$

Equating the components of G^1, G^2 and G^3 in both sides leads to:

$$N_i\sqrt{G^{ii}}\,dS=dS_i \quad i=1,2,3 \tag{3.1.5}$$

It should remembered $N_i\sqrt{G^{ii}}$ in Eq. (3.1.5) is not summed over i. So far we have not discussed about stress yet. Consider the prism $PP_1P_2P_3$ in Fig. 3.1, with the face $P_1P_2P_3$ possessing the area ΔS and the force ΔT is acting on that area. The stress can be defined by:

$$\boldsymbol{t}=\lim_{\Delta S\to 0}\frac{\Delta T}{\Delta S} \tag{3.1.6}$$

The other three faces of the prism $PP_1P_2P_3$ are in static equilibrium the stresses in each faces are $\boldsymbol{t}_1$, $\boldsymbol{t}_2$ and $\boldsymbol{t}_3$ acting on the areas dS_1, dS_2 and dS_3 in absence of inertial forces we have:

$$\boldsymbol{t}\,dS=\boldsymbol{t}_1\,dS_1+\boldsymbol{t}_2dS_2+\boldsymbol{t}_3\,dS_3 \tag{3.1.7}$$

By substituting Eq. (3.1.5) into Eq. (3.1.7) and simplifying the both sides gives:

$$\boldsymbol{t}=N_1\boldsymbol{t}_1\sqrt{G^{11}}+N_2\boldsymbol{t}_2\sqrt{G^{22}}+N_3\boldsymbol{t}_3\sqrt{G^{33}} \tag{3.1.8}$$

The important question is the directions of $\boldsymbol{t}_1$, $\boldsymbol{t}_2$ and $\boldsymbol{t}_3$? The N_1, N_2 and N_3 are scalars and are the components of $\boldsymbol{N}$ according to Eq. (3.1.4) and can be interpreted as covariant components in Eq. (3.1.8). Every term $\boldsymbol{t}_i\sqrt{G^{ii}} \quad i=1,2,3$ in Eq. (3.1.8) have components in directions G_1, G_2 and G_3 therefore they can be expanded in terms of true stress tensor τ^{ij} the first index i represents the face $\alpha_i=C \quad i=1,2,3$ of the prism $PP_1P_2P_3$ and the index j represents the direction of stress G_j $j=1,2,3$ such that the expression is:

$$t_i\sqrt{G^{ii}} = \tau^{i1}G_1 + \tau^{i2}G_2 + \tau^{i3}G_3 = \tau^{ij}G_j \quad i,j=1,2,3 \tag{3.1.9a}$$

The τ^{ij} is a contra-variant tensor expressed in a covariant base vector. Green and Zerna, stated that since N_1, N_2 and N_3 are (covariant vectors) we should express (3.1.9a) in terms of covariant base vectors. This is cannot be a true statement. The real reason is that the base vectors G_j $j=1,2,3$ lie in the surfaces $\alpha_i = C$ $i=1,2,3$ and the stress tensor cannot be defined on the base vectors that are not in those surfaces.

However, if we decide to express $t_i\sqrt{G^{ii}}$ $i=1,2,3$ in terms of G^1, G^2 and G^3 it is also possible but the tensor will change to $\tau_j^{\,i}$ which is a mixed tensor and means that first index i still represents the faces $\alpha_i = C$ $i=1,2,3$ of the prism $PP_1P_2P_3$ and the 2nd index j represents the direction of stress G^j $j=1,2,3$ which do not lie in the surfaces. These types of tensors ($\tau_j^{\,i}$) are called mixed tensors and in this case (3.1.9a) will change to:

$$t_i\sqrt{G^{ii}} = \tau_1^iG^1 + \tau_2^iG^2 + \tau_3^iG^3 = \tau_j^{\,i}\,G^j \quad i,j=1,2,3 \tag{3.1.9b}$$

The mixed tensor $\tau_j^{\,i}$ can be written into this form:

$$\tau = \tau_j^{\,r}\,G_rG^j = \tau_1^{\,1}\,G_1G^1 + \tau_2^{\,1}\,G_1G^2 + \tau_3^{\,1}\,G_1G^3 + \tau_1^{\,2}\,G_2G^1 + \tau_2^{\,2}\,G_2G^2 + \tau_3^{\,2}\,G_2G^3$$
$$+\,\tau_1^{\,3}\,G_3G^1 + \tau_2^{\,3}\,G_3G^2 + \tau_3^{\,3}\,G_3G^3$$

According to Eq. (1.2.4) we can write $G_r = G_{ir}\,G^i$ and if we substitute in the above equation we have:

$$\tau = \tau_j^{\,r}\,G_rG^j = \underbrace{\tau_j^{\,r}\,G_{ir}}_{\tau_{ij}}\,G^iG^j = \tau_{ij}G^iG^j$$

From the above expression we can conclude that:

$$\tau_{ij} = \tau_j^{\,r}\,G_{ir} \tag{3.1.10}$$

It should be reminded that covariant component of stress tensor τ_{ij} that is given by Eq. (3.1.10) does not have any significant application. Only the tensors τ^i_j and τ^{ij} are related to the stress vector *t*. In order to obtain the relationship we need to substitute Eq. (3.1.9a) and Eq. (3.1.9b) into Eq. (3.1.8) and then we have:

$$\boldsymbol{t} = N_i \tau^i_j G^j = N_i \tau^{ij} G_j \quad i,j = 1,2,3 \tag{3.1.11}$$

If we define a vector $\boldsymbol{T}_i$ according to this formula:

$$\boldsymbol{T}_i = \boldsymbol{t}_i \sqrt{G G^{ii}} \tag{3.1.12a}$$

If we substitute Eqs. (3.1.9a) and (3.1.9b) into Eq. (3.1.12a) we can find two alternative formulas for $\boldsymbol{T}_i$ as follows:

$$\boldsymbol{T}_i = \sqrt{G}\, \tau^{ij} G_j \quad \boldsymbol{T}_i = \sqrt{G}\, \tau^i_j G^j \tag{3.1.12b}$$

If we substitute Eq. (3.1.12b) into Eq. (3.1.11) a new formula for the stress vector can be found, that is,

$$\boldsymbol{t} = \frac{N_1 \boldsymbol{T}_1}{\sqrt{G}} + \frac{N_2 \boldsymbol{T}_2}{\sqrt{G}} + \frac{N_3 \boldsymbol{T}_3}{\sqrt{G}} = \frac{N_i \boldsymbol{T}_i}{\sqrt{G}} \tag{3.1.13}$$

Equations (1.3.5) to Eq. (1.3.7) indicates that the infinitesimal area $d\,a_i$ over the surface $\alpha_i = C$ is $d\,a_i = \sqrt{G G^{ii}}\, d\alpha^j d\alpha^k$, then according to Eq. (3.1.11) we can write:

$$\boldsymbol{t}_i \sqrt{G G^{ii}}\, d\alpha^j d\alpha^k = \boldsymbol{T}_i\, d\alpha^j d\alpha^k \tag{3.1.14}$$

Equation (3.1.14) shows that $\boldsymbol{t}_i$ is the true stress vector act on a true area $d\,a_i = \sqrt{G G^{ii}}\, d\alpha^j d\alpha^k$, but $\boldsymbol{T}_i$ is a virtual stress vector acts on a virtual area $d\alpha^j d\alpha^k$, meanwhile τ^{ij} is true stress tensor (Green stress). It is obvious that in non-orthogonal coordinate system elements of stress tensor τ^{ij} does not represent physical value of the stress.

The physical value of stress is named $\sigma_{(ij)}$ and is easily obtainable by resolving the stress vector $\boldsymbol{t}_i$ across unit vectors in direction of covariant-based vector $G_j\ j=1,2,3$ we designate them by $E_j\ j=1,2,3$.

$$\boldsymbol{t}_i = \sigma_{(i1)}E_1 + \sigma_{(i2)}E_2 + \sigma_{(i3)}E_3 = \sigma_{(ij)}E_j \quad i,j=1,2,3$$

However, the unit vectors are $E_j = \dfrac{G_j}{\sqrt{G_{jj}}}\ j=1,2,3$ and if we substitute into the above expression we have:

$$\boldsymbol{t}_i = \sigma_{(i1)}\frac{G_1}{\sqrt{G_{11}}} + \sigma_{(i2)}\frac{G_2}{\sqrt{G_{22}}} + \sigma_{(i3)}\frac{G_3}{\sqrt{G_{33}}} = \sigma_{(ij)}\frac{G_j}{\sqrt{G_{jj}}} \quad i,j=1,2,3$$

In order to emphasize that summation over the index j should be carried out we write the above expression into this form:

$$\boldsymbol{t}_i = \sum_{j=1}^{3}\sigma_{(ij)}\frac{G_j}{\sqrt{G_{jj}}} \quad i=1,2,3 \tag{3.1.15}$$

If we substitute Eq. (3.1.15) into Eq. (3.1.9a) we have:

$$\left(\sigma_{(i1)}\frac{G_1}{\sqrt{G_{11}}} + \sigma_{(i2)}\frac{G_2}{\sqrt{G_{22}}} + \sigma_{(i3)}\frac{G_3}{\sqrt{G_{33}}}\right)\sqrt{G^{ii}} = \tau^{i1}G_1 + \tau^{i2}G_2 + \tau^{i3}G_3$$

By equating the components of the G_1, G_2 and G_3 in both sides of the equation we can find out a relationship between $\sigma_{(ij)}$ and τ^{ij} which is:

$$\sigma_{(i1)} = \tau^{i1}\sqrt{\frac{G_{11}}{G^{ii}}} \qquad \sigma_{(i2)} = \tau^{i1}\sqrt{\frac{G_{22}}{G^{ii}}} \qquad \sigma_{(i3)} = \tau^{i1}\sqrt{\frac{G_{33}}{G^{ii}}} \tag{3.1.16}$$

In Eq. (3.1.16) $\sigma_{(ij)}$ is completely different with σ_{ij} which is known as Cauchy stress and we will describe it in this section.

In classical continuum mechanics the true stress tensor is called Cauchy stress designated by σ^{kl} and the tensor is $\boldsymbol{\sigma}$. There is not any difference between τ^{ij} and σ^{kl} except the former is expressed on G_1, G_2 and G_3 while the latter expressed g_1, g_2 and g_3. Therefore, using σ^{kl} components is more convenient and τ^{ij} components are rarely used in the literature. The Green stress tensor is like this:

$$\tau = \tau^{ij}\left[G_i\right]\left[G_j\right]^T \tag{3.1.17}$$

The Eq. (3.1.17) can be transformed into g_1, g_2 and g_3 base by using the Eq. (2.2.15) which states that:

$$\left[G_i\right] = [\boldsymbol{F}][g_i] \quad \left[G_j\right]^T = \left[g_j\right]^T [\boldsymbol{F}]^T$$

If we substitute the above equations into (3.1.17) then we have:

$$\tau = \tau^{ij}[\boldsymbol{F}][g_i]\left[g_j\right]^T[\boldsymbol{F}]^T = \sigma^{kl}[g_k][g_l]^T \tag{3.1.18}$$

In covariant base $g_1 = \begin{bmatrix}1 & 0 & 0\end{bmatrix}^T$, $g_2 = \begin{bmatrix}0 & 1 & 0\end{bmatrix}^T$ and $g_3 = \begin{bmatrix}0 & 0 & 1\end{bmatrix}^T$, therefore we can write:

$$[\boldsymbol{F}][g_i] = \begin{bmatrix} F_{1i} \\ F_{2i} \\ F_{3i} \end{bmatrix} \quad \left[g_j\right]^T[\boldsymbol{F}]^T = \begin{bmatrix} F_{1j} & F_{2j} & F_{3j} \end{bmatrix} \tag{3.1.19}$$

Substituting Eq. (3.1.19) into Eq. (3.1.18) provides the following expression:

$$\begin{bmatrix} \sigma_{11} & \sigma_{12} & \sigma_{13} \\ \sigma_{21} & \sigma_{22} & \sigma_{23} \\ \sigma_{31} & \sigma_{32} & \sigma_{33} \end{bmatrix} = \tau^{ij} \begin{bmatrix} F_{1i}F_{1j} & F_{1i}F_{2j} & F_{1i}F_{3j} \\ F_{2i}F_{1j} & F_{2i}F_{2j} & F_{2i}F_{3j} \\ F_{3i}F_{1j} & F_{3i}F_{2j} & F_{3i}F_{3j} \end{bmatrix} \quad i,j = 1,2,3 \tag{3.1.20}$$

According to Eq. (3.1.20) the Cauchy stress tensor is summation of 9 matrices all in terms of Green stress tensor. The condensed indicial form of Eq. (3.1.20) is:

$$\sigma^{kl} = \tau^{ij} F_{ki} F_{lj} \tag{3.1.21}$$

It can be rewritten into this form:

$$\sigma^{kl} = \underbrace{\tau^{ij} F_{lj}}_{\tau'^{il}} F_{ki} = \tau'^{il} F_{ki} \tag{3.1.22}$$

The expression $\tau'^{il} = \tau^{ij} F_{lj}$ in Eq. (3.1.22) can be written in matrix form as follows:

$$\left[\tau'\right] = [\tau][\boldsymbol{F}]^T \tag{3.1.23}$$

The expression $\sigma^{kl} = \tau'^{il} F_{ki}$ in Eq. (3.1.22) when written in matrix form is:

$$[\sigma] = [\boldsymbol{F}]\left[\tau'\right] \tag{3.1.24}$$

Substituting Eq. (3.1.23) into Eq. (3.1.24) we can find this formula:

$$[\sigma] = [\boldsymbol{F}][\tau][\boldsymbol{F}]^T \tag{3.1.25}$$

Equation (3.1.25) relates true Cauchy stress to true Green stress.

Now we define another type of stress, which is based on the initial un-deformed situation of the solid body. According to Fig. 3.2, the stress vector is $\boldsymbol{s}$, and the normal to the surface $P_1P_2P_3$ is designated by $\boldsymbol{n}$ and its component along the contra-variant base vectors g^1, g^2 and g^3 are:

$$\boldsymbol{n} = n_1 g^1 + n_2 g^2 + n_3 g^3 \tag{3.1.26}$$

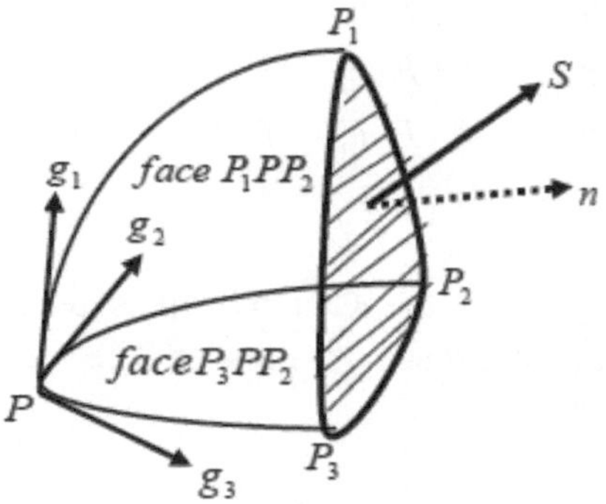

FIGURE 3.2 Stress defined in un-deformed body.

The infinitesimal areas, on the un-deformed body are ds_1, ds_2 and ds_3 the equation similar to Eq. (3.1.3) in un-deformed position is:

$$\boldsymbol{n}\,ds = \sum_{i=1}^{3} \frac{g^i ds_i}{\sqrt{g^{ii}}} \tag{3.1.27}$$

The stress vector $\boldsymbol{s}$ is also similar to Eq. (3.1.6) into this form:

$$\boldsymbol{s} = \lim_{\Delta s \to 0} \frac{\Delta T}{\Delta s} \tag{3.1.28}$$

The static equilibrium Eq. (3.1.7) will change to:

$$\boldsymbol{s}\,ds = \boldsymbol{s}_1\,ds_1 + \boldsymbol{s}_2\,ds_2 + \boldsymbol{s}_3\,ds_3 \tag{3.1.29}$$

Moreover, Eq. (3.1.8) will change to:

$$\boldsymbol{s} = n_1\boldsymbol{s}_1\sqrt{g^{11}} + n_2\boldsymbol{s}_2\sqrt{g^{22}} + n_3\boldsymbol{s}_3\sqrt{g^{33}} \tag{3.1.30}$$

In Eq. (3.1.30) the term $\boldsymbol{s}_i\sqrt{g^{ii}}$ is used to define 2nd Piola–Kirchhoff stress tensor s^{ij} over the deformed contra-variant base vectors as follows (similar to Eq. (3.1.9a)):

$$\boldsymbol{s}_i\sqrt{g^{ii}} = s^{i1}G_1 + s^{i2}G_2 + s^{i3}G_3 = s^{ij}G_j \quad i,j = 1,2,3 \tag{3.1.31}$$

Similar to Eq. (3.1.12a) and Eq. (3.1.12b) we define a vector $\boldsymbol{S}_i$ which related to the force such as:

$$\boldsymbol{S}_i = \sqrt{g}\,s^{ij}G_j \quad \boldsymbol{S}_i = \sqrt{g}\,s^i_jG^j \tag{3.1.32}$$

As Eq. (3.1.32) indicates the new components s^{ij} that represent 2nd Piola–Kirchhoff stress can be related to the Green stress τ^{ij} easily by setting $\boldsymbol{T}_i$ in Eq. (3.1.12b) and $\boldsymbol{S}_i$ in Eq. (3.1.32) equal, that is, $\boldsymbol{S}_i = \boldsymbol{T}_i$ since they are both a vector related to applied force over an area in different frames have the same value and direction therefore:

$$\sqrt{g}\,s^{ij}G_j = \sqrt{G}\,\tau^{ij}G_j \Rightarrow s^{ij} = \sqrt{\frac{G}{g}}\,\tau^{ij} \tag{3.1.33}$$

According to Eq. (1.3.8) and also Eq. (2.4.23) we had $\sqrt{\frac{G}{g}} = J = \det(\boldsymbol{F})$ and if we substitute this into Eq. (3.1.33) then we have:

$$\tau^{ij} = \frac{1}{J} s^{ij} \tag{3.1.34}$$

Substituting Eq. (3.1.34) into Eq. (3.1.25) yields to:

$$\underbrace{[\sigma]}_{\text{Cauchy Stress}} = \frac{1}{J}[\boldsymbol{F}]\underbrace{[\boldsymbol{s}]}_{\text{2nd Piola}}[\boldsymbol{F}]^T = \frac{1}{\det(\boldsymbol{F})}[\boldsymbol{F}][\boldsymbol{s}][\boldsymbol{F}]^T \tag{3.1.35}$$

Obviously the other form of (3.1.35) is:

$$\underbrace{[\boldsymbol{s}]}_{\text{2nd Piola}} = J[\boldsymbol{F}]^{-1}\underbrace{[\sigma]}_{\text{Cauchy Stress}}[\boldsymbol{F}]^{-T} = \det(\boldsymbol{F})[\boldsymbol{F}]^{-1}[\sigma][\boldsymbol{F}]^{-T} \tag{3.1.36}$$

3.2 DIFFERENT TYPES OF STRESSES AND APPLICATIONS

Stress tensor appears in equations of motion, and represents the internal forces. They will be described in the next section later. In finite element method (FEM) the component-wise inner product (Frobenius inner product), is very important because it represents virtual work of the internal forces. It can be shown with $\int_V [\sigma]:[\delta\varepsilon_v]dV$ in which $[\sigma]$ is Cauchy stress tensor and $[\delta\varepsilon_v]$ is variation of the strain tensor both tensors defined in covariant base vectors g_1, g_2 and g_3, therefore their inner product can provide exact value for the internal virtual work.

However numerical calculation of the integral $\int_V [\sigma]:[\delta\varepsilon_v]dV$ is a formidable task and instead the integral $\int_{V_0} [\boldsymbol{s}]:[\delta E_v]dV_0$ can be calculated over the initial volume V_0. In the latter integral $[\boldsymbol{s}]$ is the 2nd Piola–Kirchhoff stress and $[\delta E_v]$ variation of the strain tensor both defined on the initial un-deformed solid and can be determined easier. Both tensors $[\boldsymbol{s}]$ and $[\delta E_v]$ are not exact and correct physical stresses and strains but in FEM they can be related to $[\sigma]$ and $[\delta\varepsilon_v]$ via mapping technique.

The strain tensors $\left[\delta\varepsilon_v\right]$ and $\left[\delta E_v\right]$ can be found according the Eq. (2.2.11). If we ignore the 2nd order nonlinear term the approximate tensor $\left[\varepsilon_v\right]$ can be shown by:

$$\left[\varepsilon_v\right]\cong\begin{bmatrix}\dfrac{\partial u_1}{\partial x_1} & \dfrac{1}{2}\left(\dfrac{\partial u_1}{\partial x_2}+\dfrac{\partial u_2}{\partial x_1}\right) & \dfrac{1}{2}\left(\dfrac{\partial u_1}{\partial x_3}+\dfrac{\partial u_3}{\partial x_1}\right)\\ \dfrac{1}{2}\left(\dfrac{\partial u_2}{\partial x_1}+\dfrac{\partial u_2}{\partial x_1}\right) & \dfrac{\partial u_2}{\partial x_2} & \dfrac{1}{2}\left(\dfrac{\partial u_2}{\partial x_3}+\dfrac{\partial u_3}{\partial x_2}\right)\\ \dfrac{1}{2}\left(\dfrac{\partial u_3}{\partial x_1}+\dfrac{\partial u_1}{\partial x_3}\right) & \dfrac{1}{2}\left(\dfrac{\partial u_3}{\partial x_2}+\dfrac{\partial u_2}{\partial x_1}\right) & \dfrac{\partial u_3}{\partial x_3}\end{bmatrix} \tag{3.2.1}$$

If we compare Eq. (3.2.1) with Eq. (2.2.11) two significant differences can be found. First the nonlinear 2nd order term is ignored. The 2nd important difference is expressing $\left[\varepsilon_v\right]$ in terms of x_1, x_2 and x_3 i.e. in the deformed coordinate axis.

The Eq. (2.2.11) which gives the Green strain tensor is expressed by initial coordinates X_1, X_2 and X_3. However, the variation of strain tensor $\left[\delta\varepsilon_v\right]$ should be expressed in local coordinates x_1, x_2 and x_3 enabling the integral $\int_V[\sigma]:\left[\delta\varepsilon_v\right]dV$ to be computed. Referring to the linear terms of $\left[\varepsilon_v\right]$ the variation $\left[\delta\varepsilon_v\right]$, can be defined as follows:

$$\left[\delta\varepsilon_v\right]=\frac{1}{2}\delta\left(\left[\frac{\partial \boldsymbol{u}_v}{\partial \boldsymbol{x}}\right]+\left[\frac{\partial \boldsymbol{u}_v}{\partial \boldsymbol{x}}\right]^T\right)\Rightarrow\left[\delta\varepsilon_v\right]=\frac{1}{2}\left(\left[\frac{\partial \delta \boldsymbol{u}_v}{\partial \boldsymbol{x}}\right]+\left[\frac{\partial \delta \boldsymbol{u}_v}{\partial \boldsymbol{x}}\right]^T\right) \tag{3.2.2}$$

To write down equation (3.2.2) we have added up the matrix in Eq. (3.2.1) with its transpose and then taken the average of two. According the Eq. (2.2.6) we can find the displacement gradient matrix $\boldsymbol{D}$ by which we can find the variation of the tensor $\left[\delta E_v\right]$. Considering Green strain tensor $\left[E\right]$ can be given by Eq. (2.2.10) we can write:

$$\left[E_v\right]=\frac{1}{2}\left[\boldsymbol{D}_v+\boldsymbol{D}_v^T\right]+\frac{1}{2}\boldsymbol{D}_v\,\boldsymbol{D}_v^T \tag{3.2.3}$$

If we take the variation of $[\boldsymbol{E}]$in (3.2.3) then we have:

$$[\delta E_v] = \frac{1}{2}\left[\delta \boldsymbol{D}_v + \delta \boldsymbol{D}_v^T\right] + \frac{1}{2}\delta \boldsymbol{D}_v \boldsymbol{D}_v^T + \frac{1}{2}\boldsymbol{D}_v \delta \boldsymbol{D}_v^T + \underbrace{\frac{1}{2}\delta \boldsymbol{D}_v \delta \boldsymbol{D}_v^T}_{0} \quad (3.2.4)$$

In Eq. (3.2.4) we can ignore the last term which is the product of two variations also in Section 2.2 we said that $\boldsymbol{F} = \boldsymbol{I} + \boldsymbol{D}$ and also $\boldsymbol{F}^T = \boldsymbol{I} + \boldsymbol{D}^T$, the Eq. (3.2.4) can be written into this form:

$$[\delta E_v] = \frac{1}{2}\boldsymbol{F}^T \delta \boldsymbol{D}_v + \frac{1}{2}\delta \boldsymbol{D}_v^T \boldsymbol{F} \quad (3.2.5)$$

Now by examination of Eq. (2.2.6) we can find that:

$$\boldsymbol{D}_v = \left[\frac{\partial \boldsymbol{u}_v}{\partial \boldsymbol{X}}\right] \Rightarrow \delta \boldsymbol{D}_v = \left[\frac{\partial \delta \boldsymbol{u}_v}{\partial \boldsymbol{X}}\right] \quad (3.2.6)$$

It should be reminded that the subscript v in the formulas stands for "virtual" and in Eq. (3.2.6) the differentiation is versus the coordinate X while in Eq. (3.2.2) is versus x, the two can be related via the following manipulations:

$$\delta \boldsymbol{D}_v = \left[\frac{\partial \delta \boldsymbol{u}_v}{\partial \boldsymbol{X}}\right] = \frac{\partial \delta \boldsymbol{u}_v}{\partial \boldsymbol{x}}\frac{\partial \boldsymbol{x}}{\partial \boldsymbol{X}} = \frac{\partial \delta \boldsymbol{u}_v}{\partial \boldsymbol{x}}\boldsymbol{F} \quad (3.2.7)$$

Taking the transpose of the above matrix equation gives:

$$\delta \boldsymbol{D}_v^T = \boldsymbol{F}^T \frac{\partial \delta \mathbf{u}_v^T}{\partial \boldsymbol{x}} \quad (3.2.8)$$

If we substitute Eq. (3.2.7) and Eq. (3.2.8) into Eq. (3.2.5), then we have:

$$[\delta E_v] = \frac{1}{2}\boldsymbol{F}^T \frac{\partial \delta \boldsymbol{u}_v}{\partial \boldsymbol{x}}\boldsymbol{F} + \frac{1}{2}\boldsymbol{F}^T \frac{\partial \delta \boldsymbol{u}_v^T}{\partial \boldsymbol{x}}\boldsymbol{F}$$

By substituting Eq. (3.2.2) into the above equation we can find relationship between $[\delta E_v]$ and $[\delta \varepsilon_v]$ into this form:

$$[\delta E_v] = \boldsymbol{F}^T [\delta \varepsilon_v] \boldsymbol{F} \tag{3.2.9}$$

In calculating the integral $\int_V [\sigma]:[\delta \varepsilon_v] dV$ Eq. (3.2.9) is very important and it can be used in FEM calculations. Moreover in Section 2.2 we found that $dV = J\, dV_0$ in which $J = \det(\boldsymbol{F})$, and therefore the above integral is:

$$\int_V [\sigma]:[\delta \varepsilon_v] dV = \int_{V_0} J[\sigma]:[\delta \varepsilon_v] dV_0 \tag{3.2.10}$$

Also the integral in terms of 2nd Piola–Kirchhoff stress $\int_{V_0} [\boldsymbol{s}]:[\delta E_v] dV_0$ is:

$$\int_{V_0} [\boldsymbol{s}]:[\delta E_v] dV_0 = \int_{V_0} [\boldsymbol{s}]:\left[\boldsymbol{F}^T [\delta \varepsilon_v] \boldsymbol{F}\right] dV_0 \tag{3.2.11}$$

It is integral Eq. (3.2.11) which can be computed easily because it can be taken over the initial geometry of the body. Since $[\boldsymbol{s}] = [\boldsymbol{s}]^T$ (symmetric stress tensor) and $[\delta \varepsilon_v] = [\delta \varepsilon_v]^T$ (symmetric strain tensor) then the integral (11) is:

$$\int_{V_0} [\boldsymbol{s}]:\left[\boldsymbol{F}^T [\delta \varepsilon_v] \boldsymbol{F}\right] dV_0 = \int_{V_0} \left[\boldsymbol{F}^T [\delta \varepsilon_v]^T \boldsymbol{F}\right]:[\boldsymbol{s}]^T dV_0 \tag{3.2.12}$$

Since $\boldsymbol{a} : \boldsymbol{b} = \boldsymbol{b} : \boldsymbol{a}$ in Eq. (12) the $[\boldsymbol{s}]^T$ is moved ahead of $\boldsymbol{F}^T [\delta \varepsilon_v]^T \boldsymbol{F}$ now we defined the following terms:

$$\boldsymbol{F}^T [\delta \varepsilon_v]^T = \boldsymbol{A} \qquad \boldsymbol{F} = \boldsymbol{B} \quad [\boldsymbol{s}]^T = \boldsymbol{C}^T$$

In linear algebra we have theorem like this:

$$[\boldsymbol{AB}]:[\boldsymbol{C}]^T = [\boldsymbol{BC}]:[\boldsymbol{A}]^T \tag{3.2.13}$$

$$\left\lfloor F^T[\delta\varepsilon_v]^T F \right\rfloor:[s]^T = F\,S:[\delta\varepsilon_v]\,F = [\delta\varepsilon_v]\,F:F[S]^T \tag{3.2.14}$$

In writing the Eq. (3.2.14) we have used theorem (3.2.13) and also instead of $[s]$ in the last bracket we have used $[s]^T$. Moreover, instead of $[\delta\varepsilon_v]$ in Eq. (3.2.14) we can put $[\delta\varepsilon_v]^T$ therefore we have:

$$[\delta\varepsilon_v]\,F:F[S]^T = [\delta\varepsilon_v]^T\,F:F[S]^T \tag{3.2.15}$$

Now again define the following parameters by setting:

$$[\delta\varepsilon_v]^T = A' \qquad F = B \quad F[s]^T = C'^T \tag{3.2.16}$$

Identical expression similar to Eq. (3.2.13) can be written as:

$$\left[A'B'\right]:\left[C'\right]^T = \left[B'C'\right]:\left[A'\right]^T \tag{3.2.17}$$

According to Eq. (3.2.16) and Eq. (3.2.17) we can write:

$$[\delta\varepsilon_v]^T\,F:F[S]^T = \left(F\,[s]F^T\right):[\delta\varepsilon_v] \tag{3.2.18a}$$

If we substitute Eq. (3.2.18a) into Eq. (3.2.14) and then substitute the results in integral Eq. (3.2.12) it will change to:

$$\int_{V_0}\left[F^T[\delta\varepsilon_v]^T F\right]:[s]^T\,dV_0 = \int_{V_0}\left(F\,[s]F^T\right):[\delta\varepsilon_v]\,dV_0 \tag{3.2.18b}$$

If we compare Eq. (3.2.18b) with Eq. (3.2.12) and then with Eq. (3.2.11) we easily find out that new form of integral Eq. (3.2.11) is:

$$\int_{V_0}[s]:[\delta E_v]\,dV_0 = \int_{V_0}\left(F\,[s]F^T\right):[\delta\varepsilon_v]\,dV_0 \tag{3.2.19}$$

If we substitute Eq. (3.1.35) into Eq. (3.2.19) we can find this important relationship:

$$\int_{V_0}[s]:[\delta E_v]\,dV_0 = \int_{V_0}\left(J[\sigma]\right):[\delta\varepsilon_v]\,dV_0 \tag{3.2.20}$$

The integral $\int_V [\sigma]:[\delta\varepsilon_v]dV$ which represents the exact virtual work, is related to the integral $\int_{V_0} [s]:[\delta E_v]dV_0$ this relationship can be achieved by setting $dV = J\,dV_0$, in Eq. (3.2.19) then we have:

$$\int_V [\sigma]:[\delta\varepsilon_v]dV = \int_{V_0} [s]:[\delta E_v]dV_0 \tag{3.2.21}$$

Equation (3.2.21) shows that why the $[s]$ or the 2nd Piola–Kirchhoff stress tensor is important, because it can be replaced with Cauchy stress $[\sigma]$ according to Eq. (3.2.21). In FEM we use $[s]$ and integrate on V_0 instead of using $[\sigma]$ and integrating over V, because the latter is a formidable task.

Moreover, we can define new stress named "nominal Kirchhoff stress" $[\tau']$ given by:

$$[\tau'] = J[\sigma] \tag{3.2.22}$$

There is not any application for $[\tau']$ except it changes the Eq. (3.2.20) into new form like this:

$$\int_{V_0} [s]:[\delta E_v]dV_0 = \int_{V_0} [\tau']:[\delta\varepsilon_v]dV_0 \tag{3.2.23}$$

3.3 EQUATIONS OF MOTION AND EQUILIBRIUM

The accurate form of the equilibrium and also motion equation in large deformation can be expressed on the coordinates of the deformed body (Fig. 3.3).

For example if we write the equilibrium of the forces across along coordinate α^1 it will be in following form:

$$\underbrace{\left(F_1 + F_1\|_1\, d\alpha^1\right) - F_1}_{\text{force difference}} + \underbrace{B^1\,\rho_{df} dV}_{\text{body force}} = \underbrace{\ddot{u}^1\,\rho_{df}\, dV}_{\text{inertia force}} \tag{3.3.1}$$

The terms $F_1\|_1$ is covariant differentiation of the force component and its details are discussed in Chapter 1.3. The body force per unit mass is B and $\ddot{u}^1$

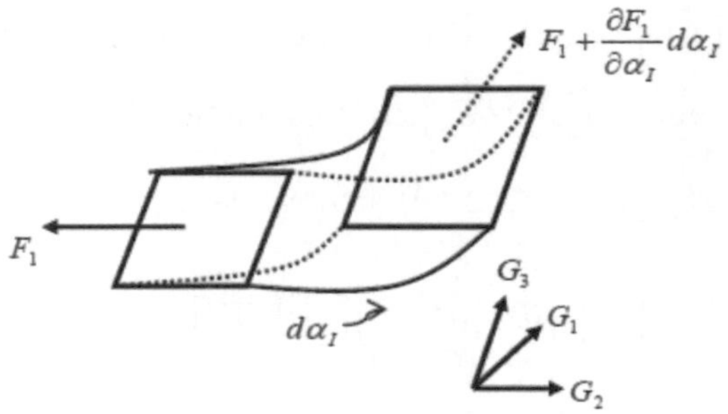

FIGURE 3.3 Equilibrium of the forces on deformed body.

is the component of acceleration along coordinate α^1. Referring to Eq. (1.3.8) we can write the volume element dV on deformed body is:

$$dV = \sqrt{G}\, d\alpha^1 d\alpha^2 d\alpha^3 \tag{3.3.2}$$

The virtual stress vector $\boldsymbol{T}_1$ is related to the force and the relationship was given by Eq. (3.1.14) as follows:

$$F_1 = T_1\, d\alpha^2 d\alpha^3 \tag{3.3.3}$$

If we substitute Eqs. (3.3.2) and (3.3.3) into Eq. (3.3.1) and simplify we have:

$$T_1\|_1 + B^1\, \rho_{df} \sqrt{G} = \ddot{u}^1\, \rho_{df} \sqrt{G} \tag{3.3.4}$$

If we generalize the above equation over $\alpha^i \quad i = 1, 2, 3$ then it will change to:

$$T_i\|_i + B^i\, \rho_{df} \sqrt{G} = \ddot{u}^i\, \rho_{df} \sqrt{G} \quad i = 1, 2, 3$$

In the above expression the summation over i does not take place. The symbol $\|$ instead of $|$ (in Chapter 1.3) used in the above equations, which represents the covariant differentiation over the deformed body. In this situation the metric tensor $\boldsymbol{g}$ will change to $\boldsymbol{G}$ and if summation over i takes place this means that we can write:

$$T_i\|_i = T_1\|_1 + T_2\|_2 + T_3\|_3$$

Then the summed up version of the above equation will be:

$$\underbrace{T_i\Big\|_i}_{\text{summation}} + \underbrace{\boldsymbol{B}}_{\text{vector}}\ \rho_{df}\sqrt{G} = \underbrace{\ddot{\boldsymbol{u}}}_{\text{vector}}\ \rho_{df}\sqrt{G} \tag{3.3.5}$$

Now if we use Eq. (3.1.12b) we can express the components of the Eq. (3.3.5) as follows:

$$T_{\mathrm{i}} = \sqrt{G}\,\tau^{ij} G_j \tag{3.3.6}$$

$$\boldsymbol{B} = B^j\, G_j \tag{3.3.7}$$

$$\ddot{\boldsymbol{u}} = \ddot{u}^j G_j \tag{3.3.8}$$

If we substitute Eqs. (3.3.6), (3.3.7) and (3.3.8) into Eq. (3.3.5) and equate the components of the G_1, G_2 and G_3 respectively then we have:

$$\tau^{ij}\Big\|_i + \rho_{df}\, B^j = \rho_{df}\, \ddot{u}^j \quad i,j = 1,2,3 \tag{3.3.9}$$

Equation (3.3.9) expresses the general equation of motion and both of the body forces and inertial forces exist in the equation. The ρ_{df} represents the density of deformed body (not initial body). The formidable task in Eq. (3.3.9) is taking covariant differentiation of the tensor components τ^{ij}. This topic was discussed in Chapter 1.3 and the Christoffel symbols were defined under Eq. (1.3.14) as follows:

$$\tau^{ij}\Big\|_k = \frac{\partial \tau^{ij}}{\partial \alpha_k} + \hat{\Gamma}^i_{km}\,\tau^{mj} + \hat{\Gamma}^j_{km}\,\tau^{ij} \tag{3.3.10}$$

In Eq. (3.3.10) the $\hat{\Gamma}$ is the Christoffel symbol which is defined for the deformed body and can be expressed similar to Eq. (1.3.15) as follows:

$$\hat{\Gamma}^k_{ji} = \frac{\partial G_j}{\partial \alpha_i}.G^k \tag{3.3.11}$$

We can summarize that the symbols $\|$ and $\hat{\Gamma}$ in this section are represented in the frame tensor $\boldsymbol{G}$, while the symbols $|$ and Γ shown in Chapter 1.3 are represented in the frame tensor $\boldsymbol{g}$.

It is obvious that Eqs. (3.3.9), (3.3.10) and (3.3.11) that together represent the equations of motion and the analytical solution of the above equations is possible only in very simple case. The alternative route is numerical solution of the above equations by using FEM. As we discussed in previous section in FEM studies it is common practice to use initial un-deformed body as frame for the tensors, that is, Eq. (3.3.9) changes to:

$$\sigma^{ij}\Big|_i + \rho B^j = \rho \ddot{u}^j \quad i,j=1,2,3 \tag{3.3.12}$$

The Eq. (3.3.12) is expressed in terms of Cauchy stress in frame of tensor $\boldsymbol{g}$ and therefore the symbol $\|$ in Eq. (3.3.9) is replaced with $|$ in Eq. (3.3.12).

Finally the issue of symmetry of the stress tensor (in large deformation) will be discussed herein. To prove the symmetry we need to write the equilibrium equation of angular momentum and the torques, which is:

$$\iint_S \underbrace{\vec{R}}_{\text{arm}} \times \underbrace{\boldsymbol{t}\,ds}_{\text{traction force}} + \iiint_V \vec{R} \times \underbrace{\rho\left(\boldsymbol{B} - \ddot{\boldsymbol{u}}\right)dV}_{\text{body \& inertia force}} = 0 \tag{3.3.13a}$$

By substituting Eq. (3.1.13) which is $\boldsymbol{t} = \dfrac{N_i \vec{T}_i}{\sqrt{G}}$ into (3.3.13a) it will change to:

$$\iint_S \vec{R} \times N_i \vec{T}_i + \iiint_V \vec{R} \times \rho\sqrt{G}\left(\boldsymbol{B} - \ddot{\boldsymbol{u}}\right)dV = 0 \tag{3.3.13b}$$

According to Eq. (3.3.13b) the moment of the traction forces is $\vec{R} \times \vec{T}_i = \vec{Q}_i$, then the surface integral is:

$$\iint_S \vec{R} \times \boldsymbol{t}\,ds = \iint_S Q_i\, N_i\, ds$$

According to the divergence theorem we can change the surface integral to volume integral and we have, $\iint_S Q_i\, N_i\, ds = \iiint_V \vec{\nabla}.\vec{Q}\,dV$ therefore the above equation is:

$$\iint_S \vec{R} \times \boldsymbol{t}\,ds = \iint_S Q_i\, N_i\, ds = \iiint_V \vec{\nabla}.\vec{Q}\,dV \tag{3.3.14}$$

The above equation is called divergence theorem and the term $\vec{\nabla}.\vec{Q}$ in Eq. (3.3.14) should be defined over the deformed base vectors such that:

$$\vec{\nabla}.\vec{Q} = Q^i\|_i = \left(\vec{R}\times\vec{T}_i\right)\|_i = \frac{\partial\vec{R}}{\partial\alpha_i}\times\vec{T}_i + \vec{R}\times T_i\|_i \tag{3.3.15}$$

In Chapter 1.1 we defined the base vectors as $\frac{\partial\vec{R}}{\partial\alpha_i} = G_i$, if we substitute Eq. (3.3.15) into Eq. (3.3.14) and then the result into Eq. (3.3.13b) then we have:

$$\iiint_V \vec{\nabla}.\vec{Q}\,dV + \iiint_V \vec{R}\times\rho\sqrt{G}\left(\boldsymbol{B}-\ddot{\boldsymbol{u}}\right)dV = 0 \tag{3.3.16a}$$

$$\iiint_V \left(G_i\times\vec{T}_i + \vec{R}\times T_i\|_i\right)dV + \iiint_V \vec{R}\times\rho\sqrt{G}\left(\boldsymbol{B}-\ddot{\boldsymbol{u}}\right)dV = 0 \tag{3.3.16b}$$

It is obvious that Eq. (3.3.16b) can be rearranged like the following:

$$\iiint_V \left(G_i\times\vec{T}_i\right)dV + \iiint_V \vec{R}\times\underbrace{\left(T_i\|_i + \rho\sqrt{G}\left(\boldsymbol{B}-\ddot{\boldsymbol{u}}\right)\right)}_{0}dV = 0 \tag{3.3.17}$$

According to equilibrium equation (3.3.5) we have $T_i\|_i + \rho\sqrt{G}\left(\boldsymbol{B}-\ddot{\boldsymbol{u}}\right) = 0$ and if we substitute this in Eq. (3.3.17) we have:

$$\iiint_V \left(G_i\times\vec{T}_i\right)dV = 0 \tag{3.3.18a}$$

From Eq. (3.3.18a) we can conclude that the integrand is also vanishes, that is,

$$G_i\times\vec{T}_i = 0 \tag{3.3.18b}$$

If we substitute Eq. (3.1.12b) into Eq. (3.3.18b) then we have:

$$G_i\times\left(\sqrt{G}\,\tau^{ij}G_j\right) = 0 \Rightarrow G_i\times\left(\tau^{ij}G_j\right) = 0 \tag{3.3.19a}$$

In Eq. (3.3.19a) if we replace the indices *i* and *j* it will change to:

$$G_j \times \left(\tau^{ji} G_i\right) = 0 \Rightarrow -G_i \times \left(\tau^{ji} G_j\right) = 0 \qquad (3.3.19b)$$

If we combine Eqs. (3.3.19b) and (3.3.19a) we can write:

$$G_i \times \left(\tau^{ij} G_j - \tau^{ji} G_j\right) = 0$$

which yields into this relationship:

$$\tau^{ij} = \tau^{ji} \qquad (3.3.20)$$

Equation (3.3.20) proves the symmetry of the Green stress tensor and subsequently the Cauchy stress and nominal Kirchhoff stress and 2nd Piola–Kirchhoff stress tensors are all symmetric.

KEYWORDS

- **divergence theorem**
- **finite element method**
- **Frobenius inner product**
- **Green stress**
- **Piola–Kirchhoff stress**
- **Cauchy stress**

CHAPTER 4

GENERAL FORM OF THE CONSTITUTIVE EQUATIONS IN SOLIDS

CONTENTS

4.1 GENERAL CONSTITUTIVE RELATIONS IN SOLID CONTINUUM

When the loading is in the range which causes small deformation in a solid continuum, the generalized Hook law expressed in tensor form is acceptable in most cases. Regardless of its tensor form it is a linear constitutive relation and are not discussed in this chapter. For detail the readers could see any continuum mechanics book, for example [11].

However, when loading causes finite deformation which could be studied in finite elasticity, hyper elasticity and finite plasticity a general constitutive law cannot be expressed by Hook's law. The general form is based on deformation energy per unit volume or w. In order to refresh the memory we can remember w in small deformation axial loading which is:

$$dw = \sigma \, d\varepsilon \tag{4.1.1}$$

$$\sigma = \frac{\partial w}{\partial \varepsilon} \tag{4.1.2}$$

In above equations σ is stress and ε is strain. In simple tension, for example, even in large strain those equations are valid. In finite deformation we can start with similar equations but instead of σ and ε we write τ^{ij} and ε_{ij} or Green stress and strain (see Chapters 2 and 3). Beforehand we define E the deformation energy per unit mass and when we refer to un-deformed body, E is:

$$w = \rho_0 \mathrm{E} \tag{4.1.3}$$

which ρ_0 is density of the un-deformed solid. We are looking for an alternative for Eq. (4.1.2) in large deformation, since stress τ^{ij} is defined on the deformed body it is not accurate to consider Eq. (4.1.3) as a candidate. Instead the energy term $\rho\mathrm{E}$ which is a quantity referred to the deformed body is an accurate candidate. Therefore, we will adopt the following alternative form for Eq. (4.1.2) for finite deformation:

$$\tau^{ij} = \frac{1}{2}\rho\left(\frac{\partial \mathrm{E}}{\partial \varepsilon_{ij}} + \frac{\partial \mathrm{E}}{\partial \varepsilon_{ji}}\right) \tag{4.1.4}$$

In Section 2.1, we defined volume change in compressible solids based on:

$$\rho_0\sqrt{g} = \rho\sqrt{G} \tag{4.1.5}$$

which g and G are third invariants of the metric tensor in initial and deformed body respectively. In Section 2.1, we defined deformation Jacobian J which changes Eq. (4.1.5) into:

$$J = \sqrt{\frac{G}{g}} \qquad \rho_0 = \rho J$$

In Section 2.4, we showed that J was expressed by I_3 the third invariant of the strain tensor and this, changes above equations into:

$$J = \sqrt{I_3} \qquad \rho = \frac{\rho_0}{\sqrt{I_3}} \tag{4.1.6}$$

By substituting Eq. (4.1.6) into Eq. (4.1.3) and the result into Eq. (4.1.4) we have:

$$\tau^{ij} = \frac{1}{2\sqrt{I_3}}\left(\frac{\partial w}{\partial \varepsilon_{ij}} + \frac{\partial w}{\partial \varepsilon_{ji}}\right) \tag{4.1.7}$$

$$w = w\left(\varepsilon_{ij}\right) \tag{4.1.8}$$

Equations (4.1.7) and (4.1.8) show that in order to drive stress stain relations, we need to express the deformation energy form of strain tensor components. It is verified by experiments that in most homogenous and isotropic materials including polymers and rubbers deformation energy does not depend on individual ε_{ij} components. Instead it depends on the invariants of the strain tensor I_1, I_2 and I_3 which yields:

$$w = w\left(I_1, I_2, I_3\right) \tag{4.1.9}$$

We rewrite Eq. (4.1.7) into following form:

$$2\tau^{ij}\sqrt{I_3} = \left(\frac{\partial w}{\partial \varepsilon_{ij}} + \frac{\partial w}{\partial \varepsilon_{ji}}\right)$$

Considering Eq. (4.1.9) and also the above equation, we can apply chain rule in differentiation to expand the above equation so that:

$$2\tau^{ij}\sqrt{I_3} = \left(\frac{\partial I_1}{\partial \varepsilon_{ij}} + \frac{\partial I_1}{\partial \varepsilon_{ji}}\right)\frac{\partial w}{\partial I_1} + \left(\frac{\partial I_2}{\partial \varepsilon_{ij}} + \frac{\partial I_2}{\partial \varepsilon_{ji}}\right)\frac{\partial w}{\partial I_2} + \left(\frac{\partial I_3}{\partial \varepsilon_{ij}} + \frac{\partial I_3}{\partial \varepsilon_{ji}}\right)\frac{\partial w}{\partial I_3} \tag{4.1.10}$$

It is now necessary to use content of Section 2.4 to evaluate the derivatives in (4.1.10) individually. We can start from Eq. (2.4.16) which is: $I_1 = g^{ij}G_{ij}$

The Green strain was: $\varepsilon_{ij} = \frac{1}{2}\left(G_{ij} - g_{ij}\right)$

That gives:

$$G_{ij} = 2\varepsilon_{ij} + g_{ij} \tag{4.1.10a}$$

From which we have:

$$I_1 = g^{ij}\left(2\varepsilon_{ij} + g_{ij}\right) \qquad \frac{\partial I_1}{\partial \varepsilon_{ij}} = 2g^{ij}$$

$$I_1 = g^{ji}\left(2\varepsilon_{ji} + g_{ji}\right) \qquad \frac{\partial I_1}{\partial \varepsilon_{ji}} = 2g^{ji} = 2g^{ij}$$

which yields:

$$\frac{\partial I_1}{\partial \varepsilon_{ij}} + \frac{\partial I_1}{\partial \varepsilon_{ji}} = 4g^{ij} \tag{4.1.11}$$

According to Eqs. (2.4.21) and (2.4.23) we can obtain partial derivative of I_3 by rewriting those equations into this form:

$$I_3 = \begin{vmatrix} g^{11} & g^{12} & g^{13} \\ g^{21} & g^{22} & g^{23} \\ g^{31} & g^{32} & g^{33} \end{vmatrix} \cdot \begin{vmatrix} G_{11} & G_{12} & G_{13} \\ G_{21} & G_{22} & G_{23} \\ G_{31} & G_{32} & G_{31} \end{vmatrix} \tag{4.1.12}$$

Considering Eq. (4.1.10a) we rewrite Eq. (4.1.12) into its new form that includes ε_{ij}:

$$I_3 = \begin{vmatrix} g^{11} & g^{12} & g^{13} \\ g^{21} & g^{22} & g^{23} \\ g^{31} & g^{32} & g^{33} \end{vmatrix} \cdot \begin{vmatrix} 2\varepsilon_{11} + g_{11} & 2\varepsilon_{12} + g_{12} & 2\varepsilon_{13} + g_{13} \\ 2\varepsilon_{21} + g_{21} & 2\varepsilon_{22} + g_{22} & 2\varepsilon_{23} + g_{23} \\ 2\varepsilon_{31} + g_{31} & 2\varepsilon_{32} + g_{32} & 2\varepsilon_{33} + g_{31} \end{vmatrix} \tag{4.1.13}$$

Imagine one wants to expand the second array in Eq. (4.1.13) to calculate the determinant of the array considering expansion by using any row and columns and then taking partial derivative relative to any ε_{ij} it is evident that they can get:

$$\frac{\partial I_3}{\partial \varepsilon_{ij}} = \frac{2I_3}{G_{ij}}$$

Exercise 4.1.1: *Prove the above expression*

Solution: By using chain rule we can write $\frac{\partial I_3}{\partial \varepsilon_{ij}} = \frac{\partial I_3}{\partial G_{ij}} \frac{\partial G_{ij}}{\partial \varepsilon_{ij}}$ and by substituting Eq. (4.1.10a) into this we have $\frac{\partial I_3}{\partial \varepsilon_{ij}} = \frac{\partial I_3}{\partial G_{ij}} \frac{\partial (2\varepsilon_{ij} + g_{ij})}{\partial \varepsilon_{ij}}$ which yields to $\frac{\partial I_3}{\partial \varepsilon_{ij}} = 2\frac{\partial I_3}{\partial G_{ij}}$

The other from Eq. (4.1.12) is $I_3 = \det(\boldsymbol{g})\det(\boldsymbol{G})$ and therefore $\frac{\partial I_3}{\partial \varepsilon_{ij}} = 2\det(\boldsymbol{g})\frac{\partial \det(\boldsymbol{G})}{\partial G_{ij}}$. Now according to Jacobi formula for the derivative of a determinant, we can write $\frac{\partial \det(\boldsymbol{G})}{\partial G_{ij}} = D^{ij}$ where D^{ij} is cofactor of the G_{ij} element, this finally leads $\frac{\partial I_3}{\partial \varepsilon_{ij}} = 2\det(\boldsymbol{g})D^{ij}$. From Eq. (1.2.7) we can write $\det(G_{ij})\det(G^{ij}) = 1$ so this results $G_{ij}D^{ij}\det(G^{ij}) = 1 \Rightarrow D^{ij} = G^{ij}\det(\boldsymbol{G})$ and by substitution $\frac{\partial I_3}{\partial \varepsilon_{ij}} = 2\det(\boldsymbol{g})\det(\boldsymbol{G})G^{ij}$ which is identical to:

$$\frac{\partial I_3}{\partial \varepsilon_{ij}} = 2I_3 G^{ij} \qquad \frac{\partial I_3}{\partial \varepsilon_{ji}} = 2I_3 G^{ji}$$

From which we can conclude that:

$$\frac{\partial I_3}{\partial \varepsilon_{ij}} + \frac{\partial I_3}{\partial \varepsilon_{ji}} = 4I_3 G^{ij} \tag{4.1.14}$$

To obtain the partial derivative of I_2 we can use the Eqs. (2.4.17c) and (2.4.17d) and rewrite I_2 into this form:

$$I_2 = \frac{1}{2}\left(I_1^2 - g^{ir}g^{js}G_{ij}G_{rs}\right)$$

The partial derivative of I_2 consist of two parts:

$$\frac{\partial I_2}{\partial \varepsilon_{ij}} = \frac{1}{2}\frac{\partial I_1^2}{\partial \varepsilon_{ij}} - \frac{1}{2}\frac{\partial g^{ir}g^{js}G_{ij}G_{rs}}{\partial \varepsilon_{ij}}$$

We can use Eq. (4.1.11) and companion equations to determine the first part as:

$$\frac{1}{2}\frac{\partial I_1^2}{\partial \varepsilon_{ij}} = I_1 \frac{\partial I_1}{\partial \varepsilon_{ij}} = g^{rs} G_{rs} \cdot 2g^{ij}$$

For the second part we substitute from Eq. (4.1.10a) for G_{ij} and G_{rs} to have:

$$-\frac{1}{2}\frac{\partial g^{ir} g^{js} G_{ij} G_{rs}}{\partial \varepsilon_{ij}} = -\frac{1}{2}\frac{\partial g^{ir} g^{js}\left(2\varepsilon_{ij} + g_{ij}\right)\left(2\varepsilon_{rs} + g_{rs}\right)}{\partial \varepsilon_{ij}} = -2g^{ir} g^{js} G_{rs}$$

By adding up the two parts we can write:

$$\frac{\partial I_2}{\partial \varepsilon_{ij}} = g^{rs} G_{rs} \cdot 2g^{ij} - 2g^{ir} g^{js} G_{rs} \qquad \frac{\partial I_2}{\partial \varepsilon_{ji}} = g^{rs} G_{rs} \cdot 2g^{ji} - 2g^{ir} g^{js} G_{rs}$$

So that: $\dfrac{\partial I_2}{\partial \varepsilon_{ij}} + \dfrac{\partial I_2}{\partial \varepsilon_{ji}} = 4\left(g^{ij} g^{rs} - g^{ir} g^{js}\right) G_{rs}$

Simplification yields:

$$\frac{\partial I_2}{\partial \varepsilon_{ij}} + \frac{\partial I_2}{\partial \varepsilon_{ji}} = 4I_1 g^{ij} - 4g^{ir} g^{js} G_{rs} \tag{4.1.15}$$

If we substitute Eqs. (4.1.1), (4.1.14) and (4.1.15) into Eq. (4.1.10) and simplify the results we can obtain the general constitutive equation as follows:

$$2\tau^{ij}\sqrt{I_3} = \left(4g^{ij}\right)\frac{\partial w}{\partial I_1} + \left(4I_1 g^{ij} - 4g^{ir} g^{js} G_{rs}\right)\frac{\partial w}{\partial I_2} + \left(4I_3 G^{ij}\right)\frac{\partial w}{\partial I_3}$$

$$\tau^{ij} = \Phi g^{ij} + \Psi B^{ij} + pG^{ij} \tag{4.1.16}$$

The scalars Φ, Ψ, p and the tensor B^{ij} in Eq. (4.1.16) are as follows:

$$\Phi = \frac{2}{\sqrt{I_3}}\frac{\partial w}{\partial I_1} \qquad \Psi = \frac{2}{\sqrt{I_3}}\frac{\partial w}{\partial I_2} \qquad p = 2\sqrt{I_3}\frac{\partial w}{\partial I_3}$$

$$B^{ij} = I_1 g^{ij} - g^{ir} g^{js} G_{rs} \tag{4.1.17}$$

Equations (4.1.16) and (4.1.17) are the constitutive equations governing any solid continuum that is under finite elastic deformation. These equations explicitly stating that, if we do not have any relationship between w and the strain invariants I_1, I_2 and I_3, then we are not able to find scalars Φ, Ψ, p, and therefore any constitutive equations. It is worth mentioning that Eqs. (4.1.16) and (4.1.17) could be found in Green and Zerna [1] and Eringen [2] without any proof. One of the missions of the book to be self-continence so that we do not miss any proof in understanding the fundamentals.

The appropriate functions for w in different materials, has been investigated by engineers and mathematicians, over the last six decade. The general form to express w in terms of I_1, I_2 and I_3 could be similar to this expression:

$$w = \sum_{p,q,r=1}^{\infty} C_{pqr} \left(I_1 - 3\right)^p \left(I_2 - 3\right)^q \left(I_3 - 1\right)^r \tag{4.1.18}$$

The importance of above expression is that in unloading condition we have $I_1, I_2 = 3$ and $I_3 = 1$ which makes the deformation energy, $w = 0$. Obviously this should be expected when the solid continuum, is unloaded. When loaded $I_1, I_2 > 3$ and $I_3 > 1$ also $w > 0$ and its value depends on stretches and strains. In the case we dealing with incompressible materials always $I_3 = 1$ and we could ignore the last term in Eq. (4.1.18) so that the modified expression is:

$$w = \sum_{p,q=1}^{\infty} C_{pq} \left(I_1 - 3\right)^p \left(I_2 - 3\right)^q \tag{4.1.19}$$

For the same reasons mentioned above, validity of Eq. (4.1.19) can also be confirmed. There are two famous models based on Eq. (4.1.19), which is acceptable in stretches up to %100, that is, $\lambda \le 2$. The first one is called Neo-Hookian model in which w depends only on first invariant $I_1 > 3$ so that in Eq. (4.1.19) we put $p = 1$ and $q = 0$. Therefore, Eq. (4.1.19) with only one term is:

$$w = C_{10} \left(I_1 - 3\right) = C_1 \left(I_1 - 3\right) \tag{4.1.20}$$

Second model which is more important is named Mooney-Rilvin w depends on first and second invariants $I_1, I_2 > 3$. The Eq. (4.1.19) in this case consists of two terms the first is with $p=1$ and $q=0$ second with $p=0$ and $q=1$ so that:

$$w = C_{10}\left(I_1 - 3\right) + C_{01}\left(I_2 - 3\right)$$

For Mooney-Rilvin type elastic solids we can write:

$$w = C_1\left(I_1 - 3\right) + C_2\left(I_2 - 3\right) \tag{4.1.21}$$

Equations (4.1.19), (4.1.20) and (4.1.21) are constitutive relations that may be considered only for incompressible materials. By those we can calculate scalars Φ, Ψ in Eq. (4.1.16) and obtain stress-deformation relationship.

Difficulty arises when we face with stretches more than %100, that is, $\lambda > 2$. This will happen in most of the rubber like materials and the Eqs. (4.1.20) and (4.1.21) are not accurate in this case. Otherwise for materials like skin, muscle, and blood vessels Eqs. (4.1.20) and (4.1.21) are acceptable. There is a model developed by Ogden [5] which is acceptable for $\lambda > 2$ and is known by his name. This model expresses the deformation energy directly in terms of stretches λ_1, λ_2 and λ_3 instead of I_1, I_2 and I_3. The general expression for Ogden model is:

$$w = \sum_{P=1}^{N} \frac{\mu_P}{\alpha_P}\left(\lambda_1^{\alpha_P} + \lambda_2^{\alpha_P} + \lambda_3^{\alpha_P} - 3\right) \tag{4.1.22}$$

For checking the validity of Eq. (4.1.22) we should mention that in unloading condition we have $\lambda_1 = \lambda_2 = \lambda_3 = 1$ which makes the deformation energy, $w = 0$. Moreover by substituting $\alpha_P = 2$ in Eq. (4.1.22) and considering only one term with $P = 1$ we can write:

$$w = \frac{\mu_1}{2}\left(\lambda_1^2 + \lambda_2^2 + \lambda_3^2 - 3\right)$$

Since in section 2-4, we had $I_1 = \lambda_1^2 + \lambda_2^2 + \lambda_3^2$ and considering $\frac{\mu_1}{2} = C_1$ the above equation becomes:

$$w = C_1\left(I_1 - 3\right)$$

which was the constitutive equation for Neo-Hookian material (see Eq. (4.1.20)).

Also in the case when we have two terms with $\alpha_1 = 2$ and $\alpha_2 = -2$ (4.1.22) is:

$$w = \frac{\mu_1}{2}\left(\lambda_1^2 + \lambda_2^2 + \lambda_3^2 - 3\right) - \frac{\mu_2}{2}\left(\lambda_1^{-2} + \lambda_2^{-2} + \lambda_3^{-2} - 3\right)$$

The above equation could be rewritten into this form:

$$w = \frac{\mu_1}{2}\left(\lambda_1^2 + \lambda_2^2 + \lambda_3^2 - 3\right) - \frac{\mu_2}{2}\left(\frac{\lambda_1^2\lambda_2^2 + \lambda_3^2\lambda_2^2 + \lambda_1^2\lambda_3^2}{\lambda_1^2\lambda_2^2\lambda_3^2} - 3\right)$$

Since from Section 2.4 we had:

$$I_1 = \lambda_1^2 + \lambda_2^2 + \lambda_3^2 \qquad I_2 = \lambda_1^2\lambda_2^2 + \lambda_3^2\lambda_2^2 + \lambda_1^2\lambda_3^2$$

For a incompressible material we have $I_3 = \lambda_1\lambda_2\lambda_3 = 1$ and therefore by substituting in the above equation w takes this form:

$$w = \frac{\mu_1}{2}\left(I_1 - 3\right) - \frac{\mu_2}{2}\left(I_2 - 3\right)$$

Now assume $C_1 = \frac{\mu_1}{2}$ and $C_2 = -\frac{\mu_2}{2}$ deformation energy w take a familiar form:

$$w = C_1\left(I_1 - 3\right) + C_2\left(I_2 - 3\right)$$

which was the constitutive equation for Mooney-Rilvin material (see Eq. (4.1.21)).

The difficulty in Ogden material model is that in its general form Eq. (4.1.22) finding an equivalent function in form of $w\left(I_1, I_2, I_3\right)$ is not possible. Therefore, we cannot use constitutive equation (4.1.16) in terms of stress.

However we should remember Ogden model is suitable for large stretches only and is not applicable for the case when large rotations and change of curvatures occurs. Considering this fact we can ignore Eq. (4.1.16) and develop a new stress-stretch relationship, which is suitable for Ogden type material. In Chapter 2 we mentioned that although stretch and strain are two different dimensionless parameters but we showed that

$d\lambda \cong d\varepsilon$. Now we can come back to the beginning of this part and write Eqs. (4.1.1) and (4.1.2) in term of $d\lambda$ instead of $d\varepsilon$ into this form:

$$dw = \sigma\, d\lambda \qquad \sigma = \frac{\partial w}{\partial \lambda} \tag{4.1.23}$$

In order to extend Eq. (4.1.23) in its general tensor form, instead of λ we use U or right stretch tensor defined in Eq. (2.3.5a) and instead of σ we have to define a new type of stress called Biot stress tensor or B. Remembering that we defined the right stretch tensor from Eq. (2.3.5a) in this form:

$$U = \lambda_1 N_1 N_1^T + \lambda_2 N_2 N_2^T + \lambda_3 N_3 N_3^T \tag{4.1.24}$$

which N_1, N_2 and N_3 were the principal directions defined on the un-deformed body. Similarly Biot stress tensor or B is also defined on N_1, N_2 and N_3 direction as:

$$B = b_1 N_1 N_1^T + b_2 N_2 N_2^T + b_3 N_3 N_3^T \tag{4.1.25}$$

From Eq. (4.1.25) it is obvious that Biot stress tensor or B is diagonal with only 3 components. Its components are defined similar to Eq. (4.1.23) by this formula.

$$b_1 = \frac{\partial w}{\partial \lambda_1} \qquad b_2 = \frac{\partial w}{\partial \lambda_2} \qquad b_3 = \frac{\partial w}{\partial \lambda_3}$$

In compact form it is:

$$b_i = \frac{\partial w}{\partial \lambda_i} \qquad i = 1,2,3 \tag{4.1.26}$$

Still we need further arguments to put forward a stress-stretch type equation for Ogden type materials. We do this in the next section. In the next part we also investigate about derivation of direct stress-deformation constitutive equations which more applicable than Eq. (4.1.16) and could be used in FEM programming.

4.2 FURTHER TYPES OF STRESSES AND CONSTITUTIVE EQUATIONS

In previous section we described the Biot stress, in this section further clarification is needed to find out the relationship between stress and stretch in Ogden model. First we study the rate of strain tensor $\dot{\varepsilon}$, since the strain ε is the tensor, its time derivative is also a tensor, that is,

$$\dot{\varepsilon} = \frac{d}{dt}(\delta\varepsilon) \tag{4.2.1}$$

In Section 3.2, we mentioned that $\delta\varepsilon$ is the variation of the local strain tensor and is referred to the deformed body (coordinate x) but not undeformed (coordinate X) now we substitute the Eq. (3.2.2) into Eq. (4.2.1) then we have:

$$\dot{\varepsilon} = \frac{d}{dt}(\delta\varepsilon) = \frac{1}{2}\frac{d}{dt}\left(\left[\frac{\partial \delta \boldsymbol{u}}{\partial \boldsymbol{x}}\right] + \left[\frac{\partial \delta \boldsymbol{u}}{\partial \boldsymbol{x}}\right]^T\right) \tag{4.2.2}$$

Now we define the strain velocity tensor $\boldsymbol{L}$ which is:

$$\boldsymbol{L} = \frac{d}{dt}\left(\frac{\partial \delta \boldsymbol{u}}{\partial \boldsymbol{x}}\right) \tag{4.2.3}$$

We define the velocity vector $\boldsymbol{v} = \dot{\boldsymbol{x}} = \frac{d}{dt}(\delta\, \boldsymbol{u})$ then Eq. (4.2.3) changes to:

$$\boldsymbol{L} = \frac{\partial \boldsymbol{v}}{\partial \boldsymbol{x}} = \frac{\partial \dot{\boldsymbol{x}}}{\partial \boldsymbol{x}} \tag{4.2.4}$$

Now we can express the strain velocity tensor $\boldsymbol{L}$ in matrix form:

$$\boldsymbol{L} = \begin{bmatrix} \frac{\partial \dot{x}_1}{\partial x_1} & \frac{\partial \dot{x}_1}{\partial x_2} & \frac{\partial \dot{x}_1}{\partial x_3} \\ \frac{\partial \dot{x}_2}{\partial x_1} & \frac{\partial \dot{x}_2}{\partial x_2} & \frac{\partial \dot{x}_2}{\partial x_3} \\ \frac{\partial \dot{x}_3}{\partial x_1} & \frac{\partial \dot{x}_3}{\partial x_2} & \frac{\partial \dot{x}_3}{\partial x_3} \end{bmatrix} \tag{4.2.5}$$

Now we can express the strain rate tensor $\dot{\varepsilon}$ in terms of tensor $\boldsymbol{L}$ as follows:

$$\dot{\varepsilon} = \frac{1}{2}\left(\boldsymbol{L} + \boldsymbol{L}^T\right) \tag{4.2.6}$$

The matrices $\boldsymbol{L}$ and $\dot{\varepsilon}$ are important and can be used in the plasticity and visco-elasticity studies, where the constitutive equations can be expressed by these matrices into this form:

$$L = \frac{\partial v}{\partial x} = \frac{1}{2}\left(L + L^T\right) - \frac{1}{2}\left(L - L^T\right) = \dot{\varepsilon} + \dot{\Omega} \tag{4.2.7}$$

In Eq. (4.2.7) we have divided the strain velocity vector $\boldsymbol{L}$ into two parts the symmetric part is the strain rate tensor $\dot{\varepsilon}$, and anti-symmetric part is $\dot{\Omega}$ called rate of rotation tensor. Herein the term $\left(L - L^T\right)$ is anti-symmetric matrix and we name it $\dot{\Omega}$ or "rotation rate tensor." Proof of the rotational property of $\dot{\Omega}$ is demonstrated later in (4.2.17). For any anti-symmetric tensor like $\dot{\Omega} = \frac{1}{2}\left(L - L^T\right)$ we have:

$$\dot{\Omega}^{\,T} = -\dot{\Omega} \tag{4.2.8}$$

In the Section 3.2 we defined the "nominal Kirchhoff stress" $\left[\tau'\right]$ by Eq. (3.2.22) and we mentioned that it is conjugate is $\left[\delta\varepsilon_v\right]$, that is, the integral $\int_{V_0}\left[\tau'\right]:\left[\delta\varepsilon_v\right]dV_0$ gives the virtual work and based on that integral we define the power as follows:

$$\dot{w} = \frac{dw}{dt} = J\sigma : \dot{\varepsilon} = \tau' : \dot{\varepsilon} \tag{4.2.9}$$

In the above equation $\dot{w}$ is the energy per unit volume (initial volume) and therefore we can use the stress $\left[\tau'\right]$. Now we can clarify the Biot stress in shadow of Eq. (4.2.9), but before that we need to define "Biot strain tensor" which is related to "right stretch tensor" like this:

$$E_b = \underbrace{U}_{\text{right stretch}} - \underbrace{I}_{\text{unit matrix}} \tag{4.2.10}$$

From Eq. (4.2.10) we can find that Biot strain tensor E_b is expressed by a diagonal matrix and does not include the shear deformations. According to Eq. (4.1.25) the Biot stresses are principal stresses also the Biot strains are the principal strains in which the rotational components are removed and from time derivative of (4.2.10) we have:

$$\dot{E}_b = \dot{U} \tag{4.2.11}$$

If we rewrite the Eq. (4.2.9) we can express it in terms of principal stresses and strains, that is, the stress tensor $\boldsymbol{B}$ and its strain conjugate E_b which is:

$$\dot{w} = \frac{dw}{dt} = \tau' : \dot{\varepsilon} = B : \dot{E}_b = B : \dot{U} \tag{4.2.12}$$

Equation (4.2.12) can be used to express a clear definition of the Biot stress in terms of power of the strain energy. Now we can refer to Section 2.2 and decompose the deformation gradient matrix to $\boldsymbol{F} = \boldsymbol{RU}$ and if we take the time derivative of this we have:

$$\boldsymbol{F} = \boldsymbol{RU} \Rightarrow \dot{\boldsymbol{F}} = \boldsymbol{R}\dot{\boldsymbol{U}} + \dot{\boldsymbol{R}}\boldsymbol{U} \tag{4.2.13}$$

Equation (4.2.13) helps to find some details about $\boldsymbol{U}$. We can rewrite the Eq. (4.2.7) and combine with Eq. (4.2.4) into this form:

$$\boldsymbol{L} = \dot{\varepsilon} + \dot{\Omega} = \frac{\partial \boldsymbol{v}}{\partial \boldsymbol{x}} = \frac{\partial \boldsymbol{v}}{\partial \boldsymbol{X}} \frac{\partial \boldsymbol{X}}{\partial \boldsymbol{x}} \tag{4.2.14}$$

Now we can say that $\boldsymbol{F} = \dfrac{\partial \boldsymbol{x}}{\partial \boldsymbol{X}}$ and therefore we can conclude that $\dot{\boldsymbol{F}} = \dfrac{\partial \boldsymbol{v}}{\partial \boldsymbol{x}}$ and also we can write $\boldsymbol{F}^{-1} = \dfrac{\partial \boldsymbol{X}}{\partial \boldsymbol{x}}$ therefore Eq. (4.2.14) can be changed to:

$$\boldsymbol{L} = \dot{\boldsymbol{F}}\boldsymbol{F}^{-1} \tag{4.2.15}$$

If we substitute Eq. (4.2.13) into (4.2.15) then we have:

$$\boldsymbol{L} = \left(\boldsymbol{R}\dot{\boldsymbol{U}} + \dot{\boldsymbol{R}}\boldsymbol{U}\right)\boldsymbol{F}^{-1} \tag{4.2.16}$$

Since we have $\boldsymbol{F} = \boldsymbol{RU}$ then we have $\boldsymbol{F}^{-1} = \boldsymbol{U}^{-1}\boldsymbol{R} = \boldsymbol{U}^{-1}\boldsymbol{R}^{T}$ and if we substitute this in Eq. (4.2.16) then we have:

$$\boldsymbol{L} = \left(\dot{\boldsymbol{R}}\boldsymbol{U} + \boldsymbol{R}\dot{\boldsymbol{U}}\right)\boldsymbol{U}^{-1}\boldsymbol{R}^{T} = \underbrace{\dot{\boldsymbol{R}}\boldsymbol{R}^{T}}_{\dot{\Omega}} + \underbrace{\boldsymbol{R}\dot{\boldsymbol{U}}\boldsymbol{U}^{-1}\boldsymbol{R}^{T}}_{\dot{\varepsilon}} \tag{4.2.17}$$

From Eq. (4.2.17) we can conclude that $\dot{\boldsymbol{R}}\boldsymbol{R}^{T}$ is rotational components of the $\boldsymbol{L}$, or $\dot{\Omega}$ (see Eq. (4.2.14)) and is anti-symmetric and $\boldsymbol{R}\dot{\boldsymbol{U}}\boldsymbol{U}^{-1}\boldsymbol{R}^{T}$ is the strain rate component of the $\boldsymbol{L}$ or $\dot{\varepsilon}$ which symmetric.

Exercise 4.2.1: *Expand* Eq. (4.2.6) *by using* Eq. (4.2.17)

Solution: Equation (4.2.6) is

$\dot{\varepsilon} = \dfrac{1}{2}\left(\underbrace{\dot{\boldsymbol{R}}\boldsymbol{R}^{T} + \boldsymbol{R}\dot{\boldsymbol{U}}\boldsymbol{U}^{-1}\boldsymbol{R}^{T}}_{\mathrm{L}} + \underbrace{\left(\dot{\boldsymbol{R}}\boldsymbol{R}^{T}\right)^{T} + \left(\boldsymbol{R}\dot{\boldsymbol{U}}\boldsymbol{U}^{-1}\boldsymbol{R}^{T}\right)^{T}}_{\mathrm{L}^{T}}\right)$ and since $\dot{\boldsymbol{R}}\boldsymbol{R}^{T}$ or $\dot{\Omega}$ is anti-symmetric then we have $\left(\dot{\boldsymbol{R}}\boldsymbol{R}^{T}\right)^{T} = -\dot{\boldsymbol{R}}\boldsymbol{R}^{T}$, then the above expression changes to $\dot{\varepsilon}=\frac{1}{2}\left(\boldsymbol{R\dot{U}U^{-1}R^{T}}+\left(\boldsymbol{R\dot{U}U^{-1}R^{T}}\right)^{T}\right)$ but the term $\left(\boldsymbol{R\dot{U}U^{-1}R^{T}}\right)^{T}=\boldsymbol{RU^{-T}\dot{U}^{T}R^{T}}$. The right stretch tensor is diagonal and therefore $\boldsymbol{U}^{-T}=\boldsymbol{U}^{-1}$ and also the $\dot{\boldsymbol{U}}^{T}=\dot{\boldsymbol{U}}$ then we have the tensor $\dot{\varepsilon}=\frac{1}{2}\left(\boldsymbol{R\dot{U}U^{-1}R^{T}}+\boldsymbol{RU^{-1}\dot{U}R^{T}}\right)$ and finally the Eq. (4.2.6) changes to:

$$\dot{\varepsilon}=\frac{1}{2}\left(\mathbf{L}+\mathbf{L}^{\mathrm{T}}\right)=\frac{1}{2}\mathbf{R}\left(\dot{\boldsymbol{U}}\boldsymbol{U}^{-1}+\boldsymbol{U}^{-1}\dot{\boldsymbol{U}}\right)\boldsymbol{R}^{T} \tag{4.2.18}$$

Now we substitute Eq. (4.2.18) into Eq. (4.2.12) then we have:

$$\dot{w}=\tau^{'}:\dot{\varepsilon}=\boldsymbol{B}:\dot{\boldsymbol{U}}=\tau^{'}:\frac{\boldsymbol{R}}{2}\left(\dot{\boldsymbol{U}}\boldsymbol{U}^{-1}+\boldsymbol{U}^{-1}\dot{\boldsymbol{U}}\right)\boldsymbol{R}^{T} \tag{4.2.19}$$

Since $\dot{\boldsymbol{U}}\boldsymbol{U}^{-1}=\boldsymbol{U}^{-1}\dot{\boldsymbol{U}}$ (both are diagonal) the Eq. (4.2.19) can be simplified to:

$$\dot{w}=\boldsymbol{B}:\dot{\boldsymbol{U}}=\tau^{'}:\boldsymbol{R\dot{U}U^{-1}R^{T}} \tag{4.2.20}$$

We use the theorem (3.2.13) $[\boldsymbol{AB}]:[\boldsymbol{C}]^{T}=[\boldsymbol{BC}]:[\boldsymbol{A}]^{T}$ several times as an exercise finally we can change the Eq. (4.2.20) into this form:

$$\dot{w}=\boldsymbol{B}:\dot{\boldsymbol{U}}=\boldsymbol{R}^{T}\tau^{'}\boldsymbol{R}\boldsymbol{U}^{-T}:\dot{\boldsymbol{U}} \tag{4.2.21}$$

From Eq. (4.2.21) we can conclude that:

$$\boldsymbol{B}=\boldsymbol{R}^{T}\tau^{'}\boldsymbol{R}\boldsymbol{U}^{-T} \tag{4.2.22a}$$

Multiplying both sides of Eq. (4.2.22a) into $\boldsymbol{U}$ then we have:

$$\boldsymbol{BU}=\boldsymbol{R}^{T}\tau^{'}\boldsymbol{R}\boldsymbol{U}^{-T}\boldsymbol{U} \tag{4.2.22b}$$

Since the right stretch matrix is diagonal then we have $\boldsymbol{U}^{-T}\boldsymbol{U}=\boldsymbol{I}$ and then we can find a relationship between Biot stress and nominal Kirchhoff stress as follows:

$$\boldsymbol{BU}=\boldsymbol{R}^{T}\tau^{'}\boldsymbol{R} \tag{4.2.23}$$

According to polar decomposition theory we have following formulas:

$$\boldsymbol{F} = \boldsymbol{R}\boldsymbol{U} \quad \boldsymbol{F}^{-1} = \boldsymbol{U}^{-1}\boldsymbol{R}^{T} \quad \boldsymbol{U}\boldsymbol{F}^{-1} = \boldsymbol{R}^{T} \quad \boldsymbol{F}^{-T}\boldsymbol{U}^{T} = \boldsymbol{R} \tag{4.2.24}$$

We can substitute $\boldsymbol{R}^{T}$ and $\boldsymbol{R}$ from Eq. (4.2.24) into Eq. (4.2.23) also considering that $\boldsymbol{U} = \boldsymbol{U}^{T}$ (diagonal matrices) we have:

$$\boldsymbol{B}\boldsymbol{U} = \boldsymbol{U}\boldsymbol{F}^{-1}\boldsymbol{\tau}'\boldsymbol{F}^{-T}\boldsymbol{U}^{T} = \boldsymbol{U}\boldsymbol{F}^{-1}\boldsymbol{\tau}'\boldsymbol{F}^{-T}\boldsymbol{U}$$

If we equate the left sides of the $\boldsymbol{U}$ in the above expression then we have:

$$\boldsymbol{B} = \boldsymbol{U}\boldsymbol{F}^{-1}\boldsymbol{\tau}'\boldsymbol{F}^{-T} \tag{4.2.25}$$

In Section 3.2 from Eq. (3.2.20) we defined 2nd Piola Kirchhoff by $\boldsymbol{S} = \boldsymbol{F}^{-1}\boldsymbol{\tau}'\boldsymbol{F}^{-T}$ and if we substitute this into Eq. (4.2.25) then we have an important expression which is:

$$\underbrace{\boldsymbol{B}}_{\text{Biot}} = \underbrace{\boldsymbol{U}}_{\text{stretch}} \underbrace{\boldsymbol{S}}_{\text{2nd Piola}} \tag{4.2.26}$$

And relates the Biot stress to 2nd Piola Kirchhoff stress and the Eq. (19) can be written in a new form like this:

$$\dot{w} = \boldsymbol{\tau}' : \dot{\varepsilon} = \boldsymbol{B} : \dot{\boldsymbol{U}} = \boldsymbol{S} : \dot{\boldsymbol{E}} \tag{4.2.27}$$

Based on the Eq. (4.2.27) we can construct a table (Table 4.1), which clearly describes the various types of stress and their conjugate strain rate tensors:

Now we return to our main discussion, which is related to the Eq. (4.2.23). In Ogden model the derivative of the stain energy are defined in terms of stretches $\frac{\partial w}{\partial \lambda_i}$ defined by Eqs. (4.1.25) and (4.1.26) and the U is also defined by Eq. (4.1.24) and since both U and B are diagonal their product is also diagonal and can be written by:

$$BU = \lambda_1 \frac{\partial w}{\partial \lambda_1} N_1 N_1^T + \lambda_2 \frac{\partial w}{\partial \lambda_2} N_2 N_2^T + \lambda_3 \frac{\partial w}{\partial \lambda_3} N_3 N_3^T \tag{4.2.28}$$

The Eq. (4.2.28) is a necessary relationship for finding properties of Ogden model. We can also use Eq. (4.2.23) to enter Nominal Kirchhoff

TABLE 4.1 Stress Strain Rate Tensors

Type of the stress tensor	Conjugate strain rate- deformation
Nominal Kirchhoff τ'	Local strain rate tensor $\dot{\varepsilon}$
2nd Piola Kirchhoff $\boldsymbol{S}$	Green strain rate tensor $\dot{E}$
Biot $\boldsymbol{B}$	Right stretch rate tensor $\dot{\mathbf{U}}$

stress τ' and Cauchy stress σ into the Ogden model. If we substitute Eqs. (4.2.23) into (4.2.28) we have:

$$\boldsymbol{R}^T\tau'\boldsymbol{R} = \lambda_1\frac{\partial w}{\partial\lambda_1}N_1N_1^T + \lambda_2\frac{\partial w}{\partial\lambda_2}N_2N_2^T + \lambda_3\frac{\partial w}{\partial\lambda_3}N_3N_3^T \qquad (4.2.29)$$

Since $\boldsymbol{R}^T = \boldsymbol{R}^{-1}$ we post-multiply both sides of Eq. (4.2.29) by $\boldsymbol{R}^T$ and we have:

$$\boldsymbol{R}^T\tau'\underbrace{\boldsymbol{R}\boldsymbol{R}^T}_{I} = \lambda_1\frac{\partial w}{\partial\lambda_1}N_1N_1^T\boldsymbol{R}^T + \lambda_2\frac{\partial w}{\partial\lambda_2}N_2N_2^T\boldsymbol{R}^T + \lambda_3\frac{\partial w}{\partial\lambda_3}N_3N_3^T\boldsymbol{R}^T$$

Also pre-multiplying the above equation by $\boldsymbol{R}$ we have:

$$\underbrace{\boldsymbol{R}\boldsymbol{R}^T}_{I}\tau'\underbrace{\boldsymbol{R}\boldsymbol{R}^T}_{I} = \lambda_1\frac{\partial w}{\partial\lambda_1}\boldsymbol{R}N_1N_1^T\boldsymbol{R}^T + \lambda_2\frac{\partial w}{\partial\lambda_2}\boldsymbol{R}N_2N_2^T\boldsymbol{R}^T + \lambda_3\frac{\partial w}{\partial\lambda_3}\boldsymbol{R}N_3N_3^T\boldsymbol{R}^T$$

which yields to this:

$$\tau' = \lambda_1\frac{\partial w}{\partial\lambda_1}\boldsymbol{R}N_1N_1^T\boldsymbol{R}^T + \lambda_2\frac{\partial w}{\partial\lambda_2}\boldsymbol{R}N_2N_2^T\boldsymbol{R}^T + \lambda_3\frac{\partial w}{\partial\lambda_3}\boldsymbol{R}N_3N_3^T\boldsymbol{R}^T \qquad (4.2.30)$$

In Eq. (2.3.28) it is shown that $\boldsymbol{n} = \boldsymbol{R}N$ and $\boldsymbol{n}^T = N^T\boldsymbol{R}^T$ if we substitute this into Eq. (4.2.30) it will change to:

$$\tau' = \lambda_1\frac{\partial w}{\partial\lambda_1}\boldsymbol{n}_1\boldsymbol{n}_1^T + \lambda_2\frac{\partial w}{\partial\lambda_2}\boldsymbol{n}_2\boldsymbol{n}_2^{TT} + \lambda_3\frac{\partial w}{\partial\lambda_3}\boldsymbol{n}_3\boldsymbol{n}_3^T \qquad (4.2.31)$$

In Section 3.2 we saw that $\tau' = J\sigma$ and therefore the Cauchy stress is:

$$\sigma = \frac{\lambda_1}{J}\frac{\partial w}{\partial \lambda_1}\boldsymbol{n}_1\boldsymbol{n}_1^T + \frac{\lambda_2}{J}\frac{\partial w}{\partial \lambda_2}\boldsymbol{n}_2\boldsymbol{n}_2^{TT} + \frac{\lambda_3}{J}\frac{\partial w}{\partial \lambda_3}\boldsymbol{n}_3\boldsymbol{n}_3^T \qquad (4.2.32)$$

It should be emphasized that Eq. (4.2.32) gives the principal Cauchy stresses with the principal components σ_1, σ_2 and σ_3 (no shear component). Equation (4.1.22) expresses the w for the Ogden model and if we take the derivative versus λ_i we have:

$$\frac{\partial w}{\partial \lambda_i} = \sum_{P=1}^{N} \mu_P \lambda_i^{\alpha_P - 1} \qquad i = 1,2,3 \qquad (4.2.33)$$

The principal components of Cauchy stress via Eq. (4.2.32) is:

$$\sigma_i = \frac{\lambda_i}{J}\frac{\partial w}{\partial \lambda_i} \qquad i = 1,2,3 \qquad (4.2.34)$$

For the incompressible materials $J = 1$ and if we substitute Eq. (4.2.33) into Eq. (4.2.34) then we have:

$$\sigma_i = \sum_{P=1}^{N} \mu_P \lambda_i^{\alpha_P} \qquad i = 1,2,3 \qquad (4.2.35)$$

Equation (4.2.35) gives direct relationship between principal Cauchy stresses and the principal stretches and has been used in FEM commercial software like ABAQUS.

4.3 PLASTICITY AND YIELD CRITERIA

The material like rubber, plastics and polymers that can undergo large deformations and still remain elastic are known as Hyper-elastic material. Unlike these materials, the metals that are main material in machine parts and structures are not Hyper-elastic but nearly rigid. Before facing large deformations, they yield and undergo plastic deformation such that if the load is removed the material does not return to its initial configuration and some permanent deformation remains. During subsequent loading and unloading events taking account of the several permanent deformations is a formidable task and FEM should be implemented for determination of the deformation history.

The plastic deformation starts from a point that yield occurs in the material. Therefore, it is necessary to discuss about yield criteria before analyzing the plastic deformation. It is important to find out about amount of the loads and stresses just when yield occurs. There are many yield criteria particularly for metals and ductile materials. In this section we discuss Von Mises yield criteria, which is also termed as J_2 yield theory. It is an important criteria which depends on the stress tensor and can be related to the loads easily, it is also applicable to large plastic deformations.

This criteria is built upon the amount of distortion energy and therefore is appropriate for studies in impact mechanics and sheet metal forming. Yielding occurs when the deformations and strains are still small and all types of the stresses like Green stress, Cauchy stress, 2nd Piola Kirchhoff stress, etc. are all identical and can be defined on the initial configuration of the body by this tensor:

$$\sigma = \begin{bmatrix} \sigma_x & \tau_{xy} & \tau_{xz} \\ \tau_{yx} & \sigma_y & \tau_{yz} \\ \tau_{zx} & \tau_{zy} & \sigma_z \end{bmatrix}$$

Which is obviously defined on the initial Cartesian coordinate system. Because of small strains in the yield onset an infinitesimal small cubic elements with dimensions *dx*, *dy* and *dz* according to Fig. 4.1, is enough for studying the stress state in the material in small deformation and strain. The stress tensor is also symmetric, that is, $\tau_{xy} = \tau_{yx}$, $\tau_{xz} = \tau_{zx}$ and also $\tau_{yz} = \tau_{zy}$.

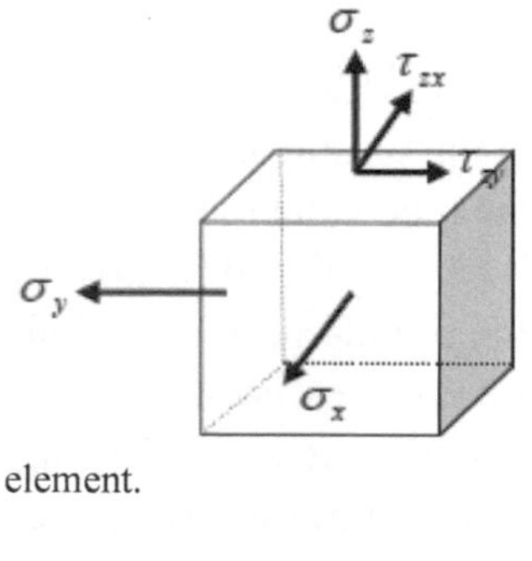

FIGURE 4.1 Infinitesimal element.

The strain energy per unit volume when stress tensor is σ can be found by:

$$U=\frac{1}{2}\sigma_x\varepsilon_x+\frac{1}{2}\sigma_y\varepsilon_y+\frac{1}{2}\sigma_z\varepsilon_z+\frac{1}{2}\tau_{xy}\gamma_{xy}+\frac{1}{2}\tau_{xz}\gamma_{xz}+\frac{1}{2}\tau_{yz}\gamma_{yz} \quad (4.3.1)$$

The Hook's law relates the stress and strain for isotropic material is summarized by:

$$\varepsilon_x=\frac{\sigma_x}{E}-\frac{\upsilon}{E}\left(\sigma_y+\sigma_z\right) \quad \gamma_{xy}=\frac{\tau_{xy}}{G} \quad (4.3.2a)$$

$$\varepsilon_y=\frac{\sigma_y}{E}-\frac{\upsilon}{E}\left(\sigma_x+\sigma_z\right) \quad \gamma_{xz}=\frac{\tau_{xz}}{G} \quad (4.3.2b)$$

$$\varepsilon_z=\frac{\sigma_z}{E}-\frac{\upsilon}{E}\left(\sigma_y+\sigma_x\right) \quad \gamma_{yz}=\frac{\tau_{yz}}{G} \quad (4.3.2c)$$

where in Eq. (4.3.2) the shear modulus G is:

$$G=\frac{E}{2(1+\upsilon)} \quad (4.3.3)$$

If we substitute Eqs. (4.3.2) and (4.3.1) and simplify we have:

$$U=\frac{1}{2E}\left[\sigma_x^2+\sigma_y^2+\sigma_z^2-2\upsilon\left(\sigma_x\sigma_y+\sigma_x\sigma_z+\sigma_z\sigma_y\right)\right] +\frac{(1+\upsilon)}{E}\left(\tau_{xy}^2+\tau_{xz}^2+\tau_{zy}^2\right) \quad (4.3.4)$$

Equation (4.3.4) gives the total energy, now we need to know what percentage of the above energy is related to distortion and shear. The necessary prelude is definition of the hydrostatic stress, which is designated by σ_m. According to Fig. 4.2, if σ_m is applied to all faces of infinitesimal cube it results no shear stress and the energy is acquired by pure tension or pure compression.

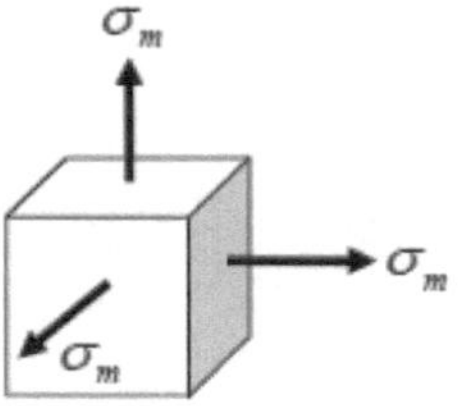

FIGURE 4.2 Hydrostatic stress over cube faces.

The hydrostatic stress is defined by:

$$\sigma_m = \frac{\sigma_x + \sigma_y + \sigma_z}{3} \tag{4.3.5}$$

The acquired strain energy U_h as a result of hydrostatic stress is:

$$U_h = \frac{1}{2}\sigma_m \varepsilon_{mx} + \frac{1}{2}\sigma_m \varepsilon_{my} + \frac{1}{2}\sigma_m \varepsilon_{mz} \tag{4.3.6}$$

The strains as a result of hydrostatic stress can be found by using Eq. (4.3.2) that represent Hook's law. The strains are:

$$\varepsilon_{mx} = \varepsilon_{my} = \varepsilon_{mz} = \frac{\sigma_m}{E} - \frac{\upsilon}{E}(\sigma_m + \sigma_m) = \frac{(1-2\upsilon)}{E}\sigma_m \tag{4.3.7}$$

If we substitute Eq. (4.3.7) into Eq. (4.3.6) and simplify then we have:

$$U_h = \frac{3(1-2\upsilon)}{2E}\sigma_m^2 \tag{4.3.8}$$

Now we substitute Eq. (4.3.5) into (Eq. 4.3.8) and simplify then we have:

$$U_h = \frac{(1-2\upsilon)}{6E}\left[\sigma_x^2 + \sigma_y^2 + \sigma_z^2 + 2\left(\sigma_x\sigma_y + \sigma_x\sigma_z + \sigma_z\sigma_y\right)\right] \tag{4.3.9}$$

The total strain energy is sum of the extensional energy U_h and distortional energy U_d, that is,:

$$U = U_h + U_d \tag{4.3.10}$$

Now we substitute Eq. (4.3.4) and Eq. (4.3.9) into Eq. (4.3.10) then after simplification we can find distortional strain energy U_d by:

$$U_d = \frac{(1+\upsilon)}{3E}\left[\sigma_x^2+\sigma_y^2+\sigma_z^2-\left(\sigma_x\sigma_y+\sigma_x\sigma_z+\sigma_z\sigma_y\right)\right] + \frac{(1+\upsilon)}{E}\left(\tau_{xy}^2+\tau_{xz}^2+\tau_{zy}^2\right) \quad (4.3.11)$$

The Von Mises equivalent stress can be defined as a uni-axial stress σ_e that can provide equal U_d. Now in Eq. (4.3.11) we set $\sigma_x = \sigma_e$ and $\sigma_y = \sigma_z = 0$ also in absence of shear stresses $\tau_{xy} = \tau_{xz} = \tau_{yz} = 0$ in Eq. (4.3.11) we have:

$$U_d = \frac{(1+\upsilon)}{3E}\sigma_e^2 \quad (4.3.12)$$

If we equate the right sides of Eqs. (4.3.11) and (4.3.12) then we have:

$$\sigma_e^2 = \left[\sigma_x^2+\sigma_y^2+\sigma_z^2-\left(\sigma_x\sigma_y+\sigma_x\sigma_z+\sigma_z\sigma_y\right)+3\left(\tau_{xy}^2+\tau_{xz}^2+\tau_{zy}^2\right)\right] \quad (4.3.13a)$$

Since from algebra we have:

$$2\left[\sigma_x^2+\sigma_y^2+\sigma_z^2-\left(\sigma_x\sigma_y+\sigma_x\sigma_z+\sigma_z\sigma_y\right)\right] = \left(\sigma_x-\sigma_y\right)^2+\left(\sigma_z-\sigma_y\right)^2 + \left(\sigma_x-\sigma_z\right)^2$$

If we substitute the above expression in Eq. (4.3.13a) and then take the square root we have:

$$\sigma_e = \sqrt{\frac{\left(\sigma_x-\sigma_y\right)^2+\left(\sigma_z-\sigma_y\right)^2+\left(\sigma_x-\sigma_z\right)^2+6\left(\tau_{xy}^2+\tau_{xz}^2+\tau_{zy}^2\right)}{2}} \quad (4.3.13b)$$

In order to express the Von Mises stress in Eq. (4.3.13b) into a simpler form we define new type of stress named "stress deviator." Stress deviator tensor is the same as stress tensor σ, except the hydrostatic stress is taken away from diagonal elements, that is,

$$\sigma' = \begin{bmatrix} \sigma_x - \sigma_m & \tau_{xy} & \tau_{xz} \\ \tau_{yx} & \sigma_y - \sigma_m & \tau_{yz} \\ \tau_{zx} & \tau_{zy} & \sigma_z - \sigma_m \end{bmatrix} \tag{4.3.14a}$$

Now by using the definition in Eq. (4.3.5) we can find the stress deviators like this:

$$\sigma_x^{'} = \sigma_x - \sigma_m = \sigma_x - \frac{\sigma_x + \sigma_y + \sigma_z}{3} = \frac{2\sigma_x - \sigma_y - \sigma_z}{3} \tag{4.3.14b}$$

Similar to Eq. (4.3.14b), we can write the other two deviator, s, that is,

$$\sigma_y^{'} = \frac{2\sigma_y - \sigma_x - \sigma_z}{3} \tag{4.3.14c}$$

$$\sigma_z^{'} = \frac{2\sigma_z - \sigma_x - \sigma_y}{3} \tag{4.3.14d}$$

Squaring Eqs. (4.3.14b), (4.3.14c) and (4.3.14d) and adding up and using elementary algebra for simplification we have:

$$\sigma_x^{'2} + \sigma_y^{'2} + \sigma_z^{'2} = \frac{6\sigma_x^2 + 6\sigma_y^2 + 6\sigma_z^2 - 6\left(\sigma_x\sigma_y + \sigma_x\sigma_z + \sigma_z\sigma_y\right)}{9}$$

The above expression can be further simplified by using:

$$2\left[\sigma_x^2 + \sigma_y^2 + \sigma_z^2 - \left(\sigma_x\sigma_y + \sigma_x\sigma_z + \sigma_z\sigma_y\right)\right] = \left(\sigma_x - \sigma_y\right)^2 + \left(\sigma_z - \sigma_y\right)^2 + \left(\sigma_x - \sigma_z\right)^2$$

Into this form:

$$\sigma_x^{'2} + \sigma_y^{'2} + \sigma_z^{'2} = \frac{\left(\sigma_x - \sigma_y\right)^2 + \left(\sigma_z - \sigma_y\right)^2 + \left(\sigma_x - \sigma_z\right)^2}{3} \tag{4.3.15}$$

If we substitute Eq. (4.3.15) into Eq. (4.3.13b) then we have:

$$\sigma_e = \sqrt{3}\sqrt{\frac{1}{2}\left(\sigma_x^{'2} + \sigma_y^{'2} + \sigma_z^{'2}\right) + \tau_{xy}^2 + \tau_{xz}^2 + \tau_{zy}^2} \tag{4.3.16}$$

Equation (4.3.16) gives the σ_e in terms of components of the stress deviators tensor σ' in Eq. (4.3.14a). Obviously it can also be expressed by (4.3.13) as well. Now we can define Von Mises yield function $f(\sigma)$ as follows:

$$f(\sigma)=\sigma_e-\sigma_0 \tag{4.3.17a}$$

In yield function (4.3.17) the σ_0 is the uni-axial yield stress of the material. The Von Mises yield criteria states that if σ_e approaches the value of σ_0 the plastic deformation starts and the yield surface can be expressed by:

$$f(\sigma)=f(\sigma')=\sigma_e-\sigma_0=0 \tag{4.3.17b}$$

Equation (4.3.17b) is the yield surface, which also can be expressed in terms of $f(\sigma')$ or stress deviator function by help of Eq. (4.3.16). In fact Eq. (4.3.17b) can be expressed into an alternative form which is:

$$f(\sigma')=\sqrt{\frac{3}{2}}(\sigma':\sigma')-\sigma_0=0 \tag{4.3.18a}$$

In Eq. (4.3.18a) σ' is the stress deviator tensor given by:

$$\sigma'=\begin{bmatrix}\sigma'_x & \tau_{xy} & \tau_{xz}\\ \tau_{yx} & \sigma'_y & \tau_{yz}\\ \tau_{zx} & \tau_{zy} & \sigma'_z\end{bmatrix} \tag{4.3.18b}$$

The term $\sigma':\sigma'$ is the scalar or inner product of two matrices (element by element multiplication and adding up), that is,

$$\sigma':\sigma'=\left(\sigma_x'^2+\sigma_y'^2+\sigma_z'^2\right)+2\tau_{xy}^2+2\tau_{xz}^2+2\tau_{zy}^2$$

Equations (4.3.17b) and (4.3.18a) both express the yield surface in different form. Because of the importance of the Von Mises yield surface it has been expressed in various forms all obviously are the same.

An alternative expression is via stretching the stress deviator tensor in vector like this:

$$\bar{\sigma}'^{T} - \left[\sigma_x' \quad \sigma_y' \quad \sigma_z' \quad \tau_{xy} \quad \tau_{xz} \quad \tau_{yz} \right] \tag{4.3.19}$$

According to Eq. (4.3.19) we can rewrite Eq. (4.3.18a) into new form:

$$\sqrt{\frac{3}{2}}\left(\bar{\sigma}'^{T} \boldsymbol{L}\, \bar{\sigma}'\right) - \sigma_0 = 0 \qquad \boldsymbol{L} = \begin{bmatrix} 1 & 0 & 0 & 0 & 0 & 0 \\ 0 & 1 & 0 & 0 & 0 & 0 \\ 0 & 0 & 1 & 0 & 0 & 0 \\ 0 & 0 & 0 & 2 & 0 & 0 \\ 0 & 0 & 0 & 0 & 2 & 0 \\ 0 & 0 & 0 & 0 & 0 & 2 \end{bmatrix} \tag{4.3.20}$$

KEYWORDS

- **Biot stress tensor**
- **finite elasticity**
- **finite plasticity**
- **Green stress**
- **Hook law**
- **hyper elasticity**
- **Jacobi formula**
- **Neo-Hookian model**
- **Ogden material model**

CHAPTER 5

STRESS–STRAIN RELATIONSHIP IN LARGE DEFORMATION OF SOLIDS

CONTENTS

5.1 STRESS–STRAIN RELATIONSHIP IN LARGE DEFORMATIONS

In classical elasticity a fourth order tensor defined and named "Hook tensor" which relates stress and strain tensors in small deformations. The question arises whether or not such expression exists in large deformations. We have answered to this question in Section 4.1, where the Eq. (4.1.16) expresses the relationship between the stress and deformation (not strain). In this section we try to find some relationship between stress and strain. If we combine the Eqs. (4.1.5), (4.1.6) and (4.1.7) we can easily find that:

$$\tau^{ij} = \frac{1}{2J}\left(\frac{\partial w}{\partial \varepsilon_{ij}} + \frac{\partial w}{\partial \varepsilon_{ji}}\right) \tag{5.1.1}$$

In Section 2.2, we showed that Green strain tensor ε_{ij} versus the deformation gradient tensor $\boldsymbol{F}$ can be defined via left Cauchy-Green tensor via Eq. (2.2.9), that is,

$$\left[\varepsilon_{ij}\right] = \frac{1}{2}\left(\boldsymbol{F}^T\boldsymbol{F} - \boldsymbol{I}\right) = \frac{1}{2}\left(\boldsymbol{c} - \boldsymbol{I}\right) \quad \boldsymbol{c} = \boldsymbol{F}^T\boldsymbol{F} \tag{5.1.2}$$

According to differentiation rules we have:

$$\frac{\partial w}{\partial \varepsilon_{ij}} = \frac{\partial w}{\partial \boldsymbol{c}} \frac{\partial \boldsymbol{c}}{\partial \varepsilon_{ij}}$$

From Eq. (5.1.2) we have, $\boldsymbol{c} = 2\left[\varepsilon_{ij}\right] + \boldsymbol{I}$ and therefore we can write:

$$\frac{\partial \boldsymbol{c}}{\partial \varepsilon_{ij}} = 2 \Rightarrow \frac{\partial w}{\partial \varepsilon_{ij}} = 2\frac{\partial w}{\partial \boldsymbol{c}} \tag{5.1.3}$$

Furthermore, in Section 3.1, from Eq. (3.1.34) we had:

$$\tau^{ij} = \frac{1}{J} s^{ij} \tag{5.1.4}$$

If we substitute Eq. (5.1.4) in the left side and Eq. (5.1.3) in the right side of Eq. (5.1.1) and then simplify we have:

$$\boldsymbol{s} = 2\frac{\partial w}{\partial \boldsymbol{c}} \tag{5.1.5}$$

In Eq. (5.1.5) the w is a scalar but $\boldsymbol{c}$ is a matrix, now derivative of a scalar value versus a matrix in Eq. (5.1.5) seems to be ambiguous, but in fact it means that:

$$s^{ij} = 2\frac{\partial w}{\partial c_{ij}} \tag{5.1.6}$$

Also Eq. (3.1.36) can be re-expressed by:

$$\boldsymbol{s} = \boldsymbol{F}^{-1}\left(J\sigma\right)\boldsymbol{F}^{-T} \tag{5.1.7}$$

Equation (3.2.22) was the definition of nominal Kirchhoff stress, that is,

$$\tau' = J\sigma \tag{5.1.8}$$

Now we substitute Eq. (5.1.8) into Eq. (5.1.7) and the result into Eq. (5.1.5) then we have:

$$\boldsymbol{F}^{-1}\tau'\,\boldsymbol{F}^{-T} = 2\frac{\partial w}{\partial \boldsymbol{c}}$$

If we post multiply the above expression into $\boldsymbol{F}^T$ and also pre multiply by $\boldsymbol{F}$ then we have:

$$\tau' = 2\boldsymbol{F}\frac{\partial w}{\partial \boldsymbol{c}}\boldsymbol{F}^T \tag{5.1.9}$$

Equations (5.1.5) and (5.1.9) are two equations that both relate nominal Kirchhoff stress to left Cauchy Green tensor $\boldsymbol{c}$, but is not a direct relationship. The complexity arises from this fact that we face both initial (reference) configuration and deformed configuration. The "push forward" operator $\boldsymbol{F}_{\phi *}$ relates and maps the initial configuration $d\boldsymbol{X}$ to deformed configuration $d\boldsymbol{x}$, that is,

$$d\boldsymbol{X} \underbrace{\Rightarrow}_{\text{push forward}} d\boldsymbol{x} \qquad d\boldsymbol{x} = \boldsymbol{F}_{\phi *}\, d\boldsymbol{X}$$

While the "pull back" operator $\boldsymbol{F}_{\phi}^{-1}$ relates and maps the deformed configuration $d\boldsymbol{x}$ to un-deformed configuration $d\boldsymbol{X}$, that is,

$$d\boldsymbol{x} \underbrace{\Rightarrow}_{\text{pull back}} d\boldsymbol{X} \qquad d\boldsymbol{X} = \boldsymbol{F}_{\phi}^{-1}\, d\boldsymbol{x}$$

Due to the duality of the transformations described above, we cannot discuss about one type of stress only. In reference configuration we defined 2nd Piola Kirchhoff stress $\boldsymbol{s}$ but in deformed configuration we defined nominal Kirchhoff τ'.

To overcome this difficulty a Lie derivative of the τ' is defined by Simo [4] which is related to both of the "push forward" operator ϕ_* and also "pull back" operator ϕ^* and can be displayed by symbol $\Im_v$ as follows:

$$\Im_v \tau' = \phi_* \left(\frac{D}{Dt}\left(\phi^* \tau'\right)\right) \tag{5.1.10}$$

The Eq. (5.1.10) can be interpreted by these statements:

i. $\phi^* \tau'$ using pull back operator to map τ' into reference configuration.
ii. Taking material derivative $\frac{D}{Dt}$ of the stress (will be explained).
iii. Mapping the material derivative into deformed configuration by push forward operator ϕ_*.

In large deformation the Lie derivative $\Im_v \tau'$ is the only criteria that relates the stresses in different frames. According to Eq. (5.1.7) we can say that $\phi^* \tau' = \boldsymbol{s}$ and therefore, the Eq. (5.1.10) will change to:

$$\Im_v \tau' = \phi_* \left(\frac{D\boldsymbol{s}}{Dt} \right)$$

Again according to Eq. (5.1.7) the push forward operator ϕ_* for $\frac{D\boldsymbol{s}}{Dt}$ can be achieved by post multiplying it into $\boldsymbol{F}^T$ and also pre multiply by $\boldsymbol{F}$. Therefore, the above expression will change to:

$$\Im_v \tau' = \boldsymbol{F} \frac{D\boldsymbol{s}}{Dt} \boldsymbol{F}^T \tag{5.1.11a}$$

Equation (5.1.11) relates the $\Im_v \tau'$ to the tensor $\boldsymbol{s}$ if we use Eq. (5.1.7) we can write:

$$\frac{D\boldsymbol{s}}{Dt} = \frac{D}{Dt}\left(\boldsymbol{F}^{-1} \tau' \boldsymbol{F}^{-T} \right)$$

Exercise 5.1.1: *Expand the above expression*

Solution: Using the law for derivative of multiplied terms we have:

$$\frac{D}{Dt}\left(\boldsymbol{F}^{-1} \tau' \boldsymbol{F}^{-T} \right) = \left(\dot{\boldsymbol{F}}^{-1} \tau' \boldsymbol{F}^{-T} + \boldsymbol{F}^{-1} \dot{\tau}' \boldsymbol{F}^{-T} + \boldsymbol{F}^{-1} \tau' \dot{\boldsymbol{F}}^{-T} \right)$$

By post multiply the above expression into $\boldsymbol{F}^T$ and also pre multiply by $\boldsymbol{F}$ then we have:

$$\boldsymbol{F} \frac{D\boldsymbol{s}}{Dt} \boldsymbol{F}^T = \boldsymbol{F}\left(\dot{\boldsymbol{F}}^{-1} \tau' \boldsymbol{F}^{-T} + \boldsymbol{F}^{-1} \dot{\tau}' \boldsymbol{F}^{-T} + \boldsymbol{F}^{-1} \tau' \dot{\boldsymbol{F}}^{-T} \right) \boldsymbol{F}^T$$

Since $\boldsymbol{F}\,\boldsymbol{F}^{-1} = \boldsymbol{I}$ and also $\boldsymbol{F}^{-T}\,\boldsymbol{F}^T = \boldsymbol{I}$ the above expression will change to:

$$\boldsymbol{F} \frac{D\boldsymbol{s}}{Dt} \boldsymbol{F}^T = \left(\boldsymbol{F}\,\dot{\boldsymbol{F}}^{-1} \tau' + \dot{\tau}' + \tau' \dot{\boldsymbol{F}}^{-T} \boldsymbol{F}^T \right) \tag{5.1.11b}$$

It is obvious that in the above expression the terms $\dot{\boldsymbol{F}}^{-1}$ and also $\dot{\boldsymbol{F}}^{-T}$ appears and they should be opened up so that we can find an alternative expression for Eq. (5.1.11).

Exercise 5.1.2: *Find expressions for* $\boldsymbol{F}\,\dot{\boldsymbol{F}}^{-1}$ *and* $\dot{\boldsymbol{F}}^{-T}\boldsymbol{F}^{T}$.

Solution: Obviously the dot sign in the $\dot{\boldsymbol{F}}^{-1}$ and also in $\dot{\boldsymbol{F}}^{-T}$ is the sign of material derivative and since $\boldsymbol{F}\,\boldsymbol{F}^{-1}=\boldsymbol{I}$ we can write:

$$\frac{D}{Dt}\left(\boldsymbol{F}\,\boldsymbol{F}^{-1}\right)=0 \Rightarrow \dot{\boldsymbol{F}}\,\boldsymbol{F}^{-1}+\boldsymbol{F}\,\dot{\boldsymbol{F}}^{-1}=0 \Rightarrow \dot{\boldsymbol{F}}^{-1}=-\boldsymbol{F}^{-1}\dot{\boldsymbol{F}}\,\boldsymbol{F}^{-1}$$

In Section 4.2 the Eq. (4.2.15) states that $\boldsymbol{L}=\dot{\boldsymbol{F}}\,\boldsymbol{F}^{-1}$ therefore we can conclude that:

$$\dot{\boldsymbol{F}}^{-1}=-\boldsymbol{F}^{-1}\boldsymbol{L}$$

Therefore we have:

$$\boldsymbol{F}\,\dot{\boldsymbol{F}}^{-1}=-\boldsymbol{L}$$

Also $\boldsymbol{F}^{-T}\,\boldsymbol{F}^{T}=\boldsymbol{I}$ and we can write:

$$\frac{D}{Dt}\left(\boldsymbol{F}^{T}\boldsymbol{F}^{-T}\right)=0 \Rightarrow \dot{\boldsymbol{F}}^{T}\,\boldsymbol{F}^{-T}+\boldsymbol{F}^{T}\,\dot{\boldsymbol{F}}^{-T}=0 \Rightarrow \dot{\boldsymbol{F}}^{-T}=-\boldsymbol{F}^{-T}\dot{\boldsymbol{F}}\,\boldsymbol{F}^{-T}$$

Pre-multiplying the above into $\dot{\boldsymbol{F}}^{-T}\boldsymbol{F}^{T}=-\boldsymbol{F}^{-T}\dot{\boldsymbol{F}}$, since $\boldsymbol{L}=\dot{\boldsymbol{F}}\,\boldsymbol{F}^{-1}$ therefore we have $\boldsymbol{L}^{T}=\boldsymbol{F}^{-T}\dot{\boldsymbol{F}}^{T}$ and we conclude $\dot{\boldsymbol{F}}^{-T}\boldsymbol{F}^{T}=-\boldsymbol{L}^{T}$, now we can substitute these findings into Eq. (5.1.11b) and the result into Eq. (5.1.11a) we have:

$$\Im_{v}\boldsymbol{\tau}'=\boldsymbol{F}\,\frac{D\boldsymbol{s}}{Dt}\boldsymbol{F}^{T}=\dot{\boldsymbol{\tau}}'-\boldsymbol{L}\boldsymbol{\tau}'-\boldsymbol{\tau}'\,\boldsymbol{L}^{T} \tag{5.1.12}$$

As we discussed the dot sign of $\dot{\boldsymbol{\tau}}'$ in Eq. (5.1.12) is for material derivative and will be explained later. We can rewrite Eq. (5.1.5) into this form:

$$s_{AB}=2\frac{\partial w}{\partial c_{AB}} \qquad A,B=1,2,3 \tag{5.1.13}$$

Now we can take the material derivative of $\dot{s}_{AB}$ from Eq. (5.1.13) as follows:

$$\dot{s}_{AB}=2\frac{Dc_{DC}}{Dt}\,\frac{\partial^{2}w}{\partial c_{AB}\,\partial c_{CD}} \qquad A,B,C,D=1,2,3 \tag{5.1.14}$$

The elements of the matrix $\boldsymbol{c}$ or c_{DC} have the material derivative or $\frac{Dc_{DC}}{Dt}$, and the Eq. (5.1.14) summed over $C,D=1,2,3$ consists of 9 terms.

According to Eq. (5.1.14) the elasticity tensor and also $\dot{c}_{DC}$ can be defined by:

$$C_{ABCD} = 4\frac{\partial^2 w}{\partial c_{AB}\,\partial c_{CD}} \qquad \dot{c}_{DC} = \frac{Dc_{DC}}{Dt}$$

Therefore, Eq. (5.1.14) can be displayed in an abbreviated form as:

$$\dot{s}_{AB} = C_{ABCD}\frac{1}{2}\dot{c}_{DC} \tag{5.1.15a}$$

Since Eq. (5.1.15a) contains 9 terms (summed over $C, D = 1,2,3$), it can be expressed by scalar matrix product in terms of fourth order tensor $\boldsymbol{C}$ as follows:

$$\dot{\boldsymbol{s}} = \boldsymbol{C} : \frac{1}{2}\dot{\boldsymbol{c}} \tag{5.1.15b}$$

To expand Eq. (5.1.15b) we need to examine $\dot{\boldsymbol{c}}$, therefore we use the definition $\boldsymbol{c} = \boldsymbol{F}^T\boldsymbol{F}$ (see the following exercise).

Exercise 5.1.3: *Expand* $\dot{\boldsymbol{c}}$

Solution: the material derivative is $\dot{\boldsymbol{c}} = \dot{\boldsymbol{F}}^T\boldsymbol{F} + \boldsymbol{F}^T\dot{\boldsymbol{F}}$ since we had $\boldsymbol{L} = \dot{\boldsymbol{F}}\,\boldsymbol{F}^{-1}$ then we have $\boldsymbol{L}\boldsymbol{F} = \dot{\boldsymbol{F}}\,\boldsymbol{F}^{-1}\boldsymbol{F}$ or $\dot{\boldsymbol{F}} = \boldsymbol{L}\boldsymbol{F}$ which yields to $\dot{\boldsymbol{F}}^T = \boldsymbol{F}^T\boldsymbol{L}^T$ and if we substitute in $\dot{\boldsymbol{c}}$ we have $\dot{\boldsymbol{c}} = \boldsymbol{F}^T\boldsymbol{L}\boldsymbol{F} + \boldsymbol{F}^T\boldsymbol{L}^T\boldsymbol{F}$ which can manipulated into this form:

$$\dot{\boldsymbol{c}} = 2\boldsymbol{F}^T\frac{1}{2}\left(\boldsymbol{L} + \boldsymbol{L}^T\right)\boldsymbol{F}$$

Since Eq. (4.2.6) provides $\dot{\boldsymbol{\varepsilon}} = \frac{1}{2}\left(\boldsymbol{L} + \boldsymbol{L}^T\right)$ then the above expression will change to:

$$\dot{\boldsymbol{c}} = 2\boldsymbol{F}^T\dot{\boldsymbol{\varepsilon}}\,\boldsymbol{F} \tag{5.1.16}$$

By post multiplying Eq. (5.1.15b) into $\boldsymbol{F}^T$ and also pre multiply by $\boldsymbol{F}$ we have:

$$\boldsymbol{F}\,\dot{\boldsymbol{s}}\,\boldsymbol{F}^T = \boldsymbol{F}\left(\boldsymbol{C} : \frac{1}{2}\dot{\boldsymbol{c}}\right)\boldsymbol{F}^T \tag{5.1.17}$$

If we compare the Eqs. (5.1.17) and (5.1.11a) we can write new formula for the Lie derivative: $\Im_v \tau'$ into this form:

$$\Im_v \tau' = \boldsymbol{F}\left(\boldsymbol{C} : \frac{1}{2}\dot{\boldsymbol{c}}\right)\boldsymbol{F}^T \tag{5.1.18}$$

The Eq. (5.1.11a) can be written in indicial form as:

$$\left(\Im_v \tau'\right)_{ab} = F_{aA}\, \dot{s}_{AB} F_{bB} \tag{5.1.19}$$

The Eq. (5.1.16) can be written in indicial form as:

$$\frac{1}{2}\dot{c}_{CD} = F_{cC}\, \dot{\varepsilon}_{cd}\, F_{dD} \tag{5.1.20}$$

By substituting Eq. (5.1.20) into Eq. (5.1.15a) a new expression can be found which is:

$$\dot{s}_{AB} = C_{ABCD}\, F_{cC}\, \dot{\varepsilon}_{cd}\, F_{dD} \tag{5.1.21}$$

If we substitute Eq. (5.1.21) into Eq. (5.1.19) then we have:

$$\left(\Im_v \tau'\right)_{ab} = \left(F_{aA}\, F_{bB}\, F_{cC}\, F_{dD}\, C_{ABCD}\right) \dot{\varepsilon}_{cd} \tag{5.1.22}$$

The Eq. (5.1.22) can be expressed in concise form like this:

$$\left(\Im_v \tau'\right)_{ab} = \mathrm{c}_{abcd}\, \dot{\varepsilon}_{cd} \tag{5.1.23}$$

$$\mathrm{c}_{abcd} = F_{aA}\, F_{bB}\, F_{cC}\, F_{dD}\, C_{ABCD} \tag{5.1.24}$$

Equation (5.1.24) provides the relationship between elements of fourth order tensor c_{abcd} in the deformed status with fourth order tensor C_{ABCD} in the initial un-deformed reference status. Since Eq. (5.1.23) contains 9 terms (summed over $c, d = 1, 2, 3$), it can be expressed by scalar matrix product in terms of fourth order tensor $\boldsymbol{c}$ as follows:

$$\underbrace{\Im_v \tau'}_{\text{Lie derivative}} = \underbrace{\boldsymbol{c}}_{\text{tensor deformed}} : \underbrace{\dot{\boldsymbol{\varepsilon}}}_{\text{strain rate}} \tag{5.1.25}$$

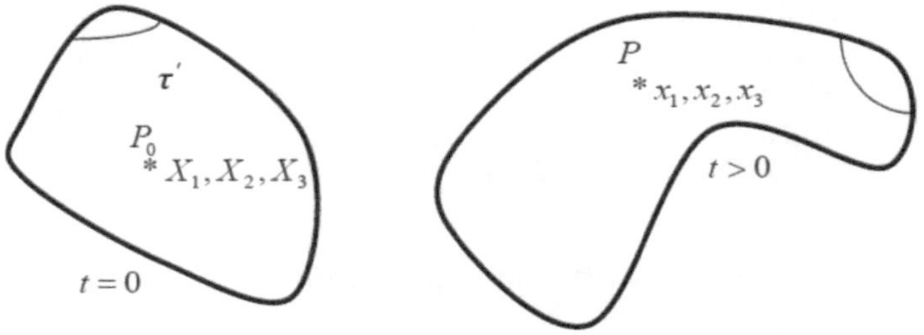

FIGURE 5.1 Initial and deformed status of a continuum.

At the end of this section we discuss about material derivative $\dot{\tau}'$ in Eq. (5.1.12). This can be explained according to Fig. 5.1, that a point P_0 in the initial un-deformed status located in the coordinate X_1, X_2, X_3 after deformation moves to the point P located in the coordinate x_1, x_2, x_3. Some quantities like temperature are scalar and depends on the location (field), while tensors like $\tau'(X_1, X_2, X_3)$ can be expressed in initial coordinate.

It will be accessible in the deformed status by $\tau'(x_1, x_2, x_3)$, the material differential $D\tau'$ does not depend to x_1, x_2, x_3 only, but it depends on time as well, that is, x_1, x_2, x_3, t if we want to follow an initial point P_0 we need to use complete differential $D\tau'$, that is,

$$D\tau' = \frac{\partial \tau'}{\partial t}dt + \frac{\partial \tau'}{\partial x_1}dx_1 + \frac{\partial \tau'}{\partial x_2}dx_2 + \frac{\partial \tau'}{\partial x_3}dx_3$$

Dividing both sides of the above expression into dt then we have:

$$\frac{D\tau'}{dt} = \frac{\partial \tau'}{\partial t} + \frac{\partial \tau'}{\partial x_1}\frac{dx_1}{dt} + \frac{\partial \tau'}{\partial x_2}\frac{dx_2}{dt} + \frac{\partial \tau'}{\partial x_3}\frac{dx_3}{dt}$$

The expression $\frac{D\tau'}{dt}$ (above) evaluates the material derivative of stress τ' and follows the point P_0 on the initial un-deformed body. It can be expanded into this form:

$$\dot{\tau}' = \frac{\partial \tau'}{\partial t} + \left(\frac{\partial \tau'}{\partial x_1}\vec{e}_1 + \frac{\partial \tau'}{\partial x_2}\vec{e}_2 + \frac{\partial \tau'}{\partial x_3}\vec{e}_3 \right) \cdot \left(\frac{dx_1}{dt}\vec{e}_1 + \frac{dx_2}{dt}\vec{e}_2 + \frac{dx_3}{dt}\vec{e}_3 \right)$$

It can also be expressed in an abbreviated form like this:

$$\dot{\tau}' = \frac{\partial \tau'}{\partial t} + \vec{\nabla}\tau' \cdot \vec{v} \tag{5.1.26}$$

In Eq. (5.1.26) $\vec{v} = \frac{dx_1}{dt}\vec{e}_1 + \frac{dx_2}{dt}\vec{e}_2 + \frac{dx_3}{dt}\vec{e}_3$ is the velocity vector of the material point and $\vec{\nabla}\tau' = \frac{\partial \tau'}{\partial x_1}\vec{e}_1 + \frac{\partial \tau'}{\partial x_2}\vec{e}_2 + \frac{\partial \tau'}{\partial x_3}\vec{e}_3$ is the gradient of the elements of the stress tensor τ'. The Eq. (5.1.26) is 9-field equation and each field is one component of the stress tensor τ'.

5.2 CONSTITUTIVE EQUATIONS FOR ELASTIC AND PLASTIC MATERIALS

In Section 4.2, we discussed about constitutive equations in elastic materials and in Section 4.3, we discussed plastic yield criteria in which Von Mises yield theory explained, where it was mentioned that formulation of the yield criteria in terms stress deviator components is much easier. However, in previous section we demonstrated that in large deformation the general constitutive equation is $\Im_v \tau' = \boldsymbol{c} : \dot{\boldsymbol{\varepsilon}}$ such that its implementation in FEM studies is a formidable task. In this section we try to derive some simple forms of the constitutive equations is that valid for elastic materials and in the next section we discuss about the constitutive equations in elastic-plastic materials.

In this section all the equations will be valid for large deformations, and like the Section 4.3 that we introduced stress deviator, we defined deviatory components of the strains and stretches. The invariants of the stretch tensor we discussed in Section 2.4, and the expressions in terms of both left and right Cauchy Green stretch tensors (**b** and **c**), that is,

$$\det\left(\boldsymbol{c} - \lambda_p^2 \boldsymbol{I}\right) = \det\left(\boldsymbol{b} - \lambda_p^2 \boldsymbol{I}\right) = 0 \tag{5.2.1}$$

The expansion of the above determinant gives:

$$\lambda_p^6 - I_1\lambda_p^4 + I_2\lambda_p^2 - I_3 = 0 \tag{5.2.2}$$

From Eq. (2.4.14) the 1st invariant is:

$$I_1 = \lambda_1^2 + \lambda_2^2 + \lambda_3^2 = c_{11} + c_{22} + c_{33} = \mathrm{tr}(\boldsymbol{c}) = \boldsymbol{I} : \boldsymbol{c} \tag{5.2.3}$$

From Eq. (2.4.17a, b) the 2nd invariant is:

$$I_2 = \lambda_1^2\lambda_2^2 + \lambda_3^2\lambda_2^2 + \lambda_1^2\lambda_3^2 = c_{11}c_{22} + c_{11}c_{33} + c_{33}c_{22} - c_{12}c_{21} - c_{13}c_{31} - c_{23}c_{32} \tag{5.2.4a}$$

$$I_2 = \frac{1}{2}\left(I_1^2 - \mathrm{tr}(\boldsymbol{c}\,\boldsymbol{c})\right) = \frac{1}{2}\left(I_1^2 - \boldsymbol{I} : \boldsymbol{c}^2\right) \tag{5.2.4b}$$

From Eq. (2.4.19a) the 3rd invariant is:

$$I_3 = \det(\boldsymbol{c})$$

Exercise 5.2.1: *find* $\dfrac{\partial I_3}{\partial \boldsymbol{c}}$

Solution: From linear algebra we have $\det(\boldsymbol{c})\boldsymbol{c}^{-1} = \mathrm{adj}(\boldsymbol{c})$ or $\det(\boldsymbol{c}) = \boldsymbol{c}\,\mathrm{adj}(\boldsymbol{c})$ if we take the derivative of this versus $\boldsymbol{c}$ then we have:

$$\det(\boldsymbol{c}) = \boldsymbol{c}\,\mathrm{adj}(\boldsymbol{c}) \Rightarrow \frac{\partial \det(\boldsymbol{c})}{\partial \boldsymbol{c}} = \mathrm{adj}(\boldsymbol{c})$$

Since $\mathrm{adj}(\boldsymbol{c}) = \det(\boldsymbol{c})\boldsymbol{c}^{-1} = I_3\,\boldsymbol{c}^{-1}$ then it is obvious that:

$$\frac{\partial I_3}{\partial \boldsymbol{c}} = I_3\,\boldsymbol{c}^{-1} \tag{5.2.5}$$

Similarly we can find $\dfrac{\partial I_1}{\partial \boldsymbol{c}}$ by using Eq. (5.2.3), that is,

$$\frac{\partial I_1}{\partial \boldsymbol{c}} = \frac{\partial}{\partial \boldsymbol{c}}(\boldsymbol{I} : \boldsymbol{c}) = \boldsymbol{I} \tag{5.2.6}$$

also we can find $\dfrac{\partial I_2}{\partial \boldsymbol{c}}$ by using Eq. (5.2.4b), that is,

$$\frac{\partial I_2}{\partial \boldsymbol{c}} = \frac{\partial\left(\frac{1}{2}I_1^2\right)}{\partial I_1}\frac{\partial I_1}{\partial \boldsymbol{c}} - \frac{1}{2}\frac{\partial}{\partial \boldsymbol{c}}\left(\boldsymbol{I} : \boldsymbol{c}^2\right) = I_1\,\boldsymbol{I} - \boldsymbol{c} \tag{5.2.7}$$

In Section 4.3, the stress tensor splat into two parts (hydrostatic and deviator), in large deformation though the deformation gradient matrix $\boldsymbol{F}$ can be decomposed into two parts the deviatory part is $\bar{\boldsymbol{F}}$ and the hydrostatic part is $J^{-1/3}$, that is,

$$\boldsymbol{F} = J^{1/3}\bar{\boldsymbol{F}}$$

The above expression yields to:

$$\bar{\boldsymbol{F}} = J^{-1/3}\boldsymbol{F} = I_3^{-1/6}\boldsymbol{F} \tag{5.2.8}$$

Regarding deviatory part $\bar{\boldsymbol{F}}$, two new tensors $\bar{\boldsymbol{c}}$ and $\bar{\boldsymbol{b}}$ also can be defined by using Eq. (5.2.8), that is,

$$\bar{\boldsymbol{c}} = \bar{\boldsymbol{F}}^T\bar{\boldsymbol{F}} = J^{-2/3}\boldsymbol{c} = I_3^{-1/3}\boldsymbol{c} \tag{5.2.9}$$

$$\bar{\boldsymbol{b}} = \bar{\boldsymbol{F}}\bar{\boldsymbol{F}}^T = J^{-2/3}\boldsymbol{b} = I_3^{-1/3}\boldsymbol{b} \tag{5.2.10}$$

The stretch invariant I_1, I_2 and I_3 are defined according to $\boldsymbol{c}$, similarly the invariants $\bar{I}_1, \bar{I}_2$ and $\bar{I}_3$ can be defined based on $\bar{\boldsymbol{c}}$, that is,

$$\bar{I}_1 = I_1 I_3^{-1/3} = I_1 J^{-2/3} \tag{5.2.11}$$

$$\bar{I}_2 = I_2 I_3^{-2/3} = I_2 J^{-4/3} \tag{5.2.12}$$

The bar sign $\bar{I}_1, \bar{I}_2$ and $\bar{I}_3$ stands for the deviatory components. An important property of deviatory $\bar{\boldsymbol{F}}$ can be found by using the Eq. (5.2.8) such that:

$$\boldsymbol{F} = J^{1/3}\bar{\boldsymbol{F}} \Rightarrow \det(\boldsymbol{F}) = \left(J^{1/3}\right)^3 \det\left(\bar{\boldsymbol{F}}\right)$$

Since we have $\det(\boldsymbol{F}) = J$, the above expression yields to:

$$\det\left(\bar{\boldsymbol{F}}\right) = 1 \tag{5.2.13}$$

Interpretation of Eq. (5.2.13) is that $\bar{\boldsymbol{F}}$ or deviatory components deals with incompressible materials where there is not volume change as a result of pressure. Since plastic deformation does not cause volume change, the matrix $\bar{\boldsymbol{F}}$ is appropriate for studying plastic deformation of the materials.

The Mooney–Rivlin, material model that defined in Section 4.1, in terms of I_1, I_2 now can be modified in terms of $\overline{I}_1, \overline{I}_2$ as follows:

$$w = w_d + w_b = \underbrace{C_1\left(\overline{I}_1 - 3\right) + C_2\left(\overline{I}_2 - 3\right)}_{w_d \text{ deviator part}} + \underbrace{\frac{1}{2}K\left(J-1\right)^2}_{w_b \text{ bulk part}} \tag{5.2.14}$$

The Eq. (5.2.14) is based on the expression $\overline{\lambda} = J^{-1/3}\lambda$, such that under pure pressure $\overline{\lambda}_1 = \overline{\lambda}_2 = \overline{\lambda}_3$ and according to Eq. (5.2.13) we have $\overline{\lambda}_1\overline{\lambda}_2\overline{\lambda}_3 = 1$. According to Eq. (5.2.14) under pure pressure the strain energy is $w = w_b$ and $w_d = 0$, and therefore we have:

$$\overline{I}_1\left(\overline{\lambda}_1, \overline{\lambda}_2, \overline{\lambda}_3\right) = 3 \quad \overline{I}_2\left(\overline{\lambda}_1, \overline{\lambda}_2, \overline{\lambda}_3\right) = 3$$

For incompressible material $w_d = 0$ and $w_b = \frac{1}{2}K\left(J-1\right)^2$ is the energy required for the compression. Therefore, Eq. ((5.2.14)) provides the total energy w, and according to Eq. (5.1.5) can be written by:

$$\boldsymbol{s} = 2\frac{\partial w\left(I_1, I_2, I_3\right)}{\partial \boldsymbol{c}} \tag{5.2.15}$$

By using chain rule in differentiation (5.2.15) can be expanded as follows:

$$\boldsymbol{s} = 2\left(\frac{\partial w}{\partial I_1}\frac{\partial I_1}{\partial \boldsymbol{c}} + \frac{\partial w}{\partial I_2}\frac{\partial I_2}{\partial \boldsymbol{c}} + \frac{\partial w}{\partial I_3}\frac{\partial I_3}{\partial \boldsymbol{c}}\right) \tag{5.2.16}$$

According to Eq. (5.2.14) we can write $\frac{\partial w}{\partial \overline{I}_1} = C_1$ and also $\frac{\partial w}{\partial \overline{I}_2} = C_2$, also from Eq. (5.2.11) we have $\frac{\partial \overline{I}_1}{\partial I_1} = I_3^{-1/3}$ and from Eq. (5.2.12) we have $\frac{\partial \overline{I}_2}{\partial I_2} = I_3^{-2/3}$. Followed by a change of variable we can write these formulas:

$$\frac{\partial w}{\partial I_1} = \frac{\partial w}{\partial \overline{I}_1}\frac{\partial \overline{I}_1}{\partial I_1} = C_1 I_3^{-1/3} \tag{5.2.17}$$

$$\frac{\partial w}{\partial I_2}=\frac{\partial w}{\partial \bar{I}_2}\frac{\partial \bar{I}_2}{\partial I_2}=C_2\, I_3^{-2/3} \tag{5.2.18}$$

$$\frac{\partial w}{\partial I_3}=\frac{\partial w_d}{\partial \bar{I}_1}\frac{\partial \bar{I}_1}{\partial I_3}+\frac{\partial w_d}{\partial \bar{I}_2}\frac{\partial \bar{I}_2}{\partial I_3}+\frac{\partial w_b}{\partial I_3} \tag{5.2.19}$$

From Eq. (5.2.14) we can write:

$$w_d=C_1\left(\bar{I}_1-3\right)+C_2\left(\bar{I}_2-3\right) \qquad w_b=\frac{1}{2}K\left(\sqrt{I_3}-1\right)^2$$

From Eqs. (5.2.11) and (5.2.12) we can write:

$$\frac{\partial \bar{I}_1}{\partial I_3}=-\frac{1}{3}I_1\, I_3^{-4/3} \qquad \frac{\partial \bar{I}_2}{\partial I_3}=-\frac{2}{3}I_2\, I_3^{-5/3} \tag{5.2.20a}$$

$$\frac{\partial w_b}{\partial I_3}=K\frac{1}{2\sqrt{I_3}}\left(\sqrt{I_3}-1\right)=\frac{K(J-1)}{2J} \tag{5.2.20b}$$

$$\frac{\partial w_d}{\partial \bar{I}_1}=C_1 \qquad \frac{\partial w_d}{\partial \bar{I}_2}=C_2 \tag{5.2.20c}$$

If we substitute Eqs. (5.2.20a), (5.2.20b) and (5.2.20c) into Eq. (5.2.19) we have:

$$\frac{\partial w}{\partial I_3}=-\frac{1}{3}C_1\, I_1\, I_3^{-4/3}-\frac{2}{3}C_2\, I_1\, I_3^{-5/3}+\frac{K(J-1)}{2J} \tag{5.2.21}$$

Before we derive a general equation we need to clarify K (bulk modulus) parameter in Eq. (5.2.20). In a situation where we have pure pressure applied to the solid, in this case the stretches in all directions are equal, that is, $\lambda_1=\lambda_2=\lambda_3=\lambda$ and the stretch invariants can be found by Eqs. (5.2.3), (5.2.4a), (5.2.11) and (5.2.12) are:

$$I_1=3\lambda^2 \qquad I_2=3\lambda^4 \qquad I_3=\lambda^6$$

$$J=\lambda^3 \qquad \bar{I}_1=3 \qquad \bar{I}_2=3 \qquad w_d=0$$

Moreover, $\dfrac{\partial I_3}{\partial \boldsymbol{c}} = J^2 \boldsymbol{c}^{-1}$ can be found from Eq. (5.2.5) such that Eq. (5.2.15) can be written by:

$$\boldsymbol{s} = 2\frac{\partial w}{\partial \boldsymbol{c}} = 2\frac{\partial w_b}{\partial \boldsymbol{c}} = 2\frac{\partial w_b}{\partial I_3}\frac{\partial I_3}{\partial \boldsymbol{c}} = 2\frac{K(J-1)}{2J}\left(J^2 \boldsymbol{c}^{-1}\right) = KJ(J-1)\boldsymbol{c}^{-1}$$

which can be simplified into:

$$\boldsymbol{s} = KJ(J-1)\boldsymbol{c}^{-1} \tag{5.2.22}$$

Equation (5.2.22) gives relationship between $\boldsymbol{s}$ (2nd Piola stress) with $\boldsymbol{c}^{-1}$ in case of hydrostatic pressure loading. In this particular case we have:

$$\boldsymbol{c} = \boldsymbol{F}^T \boldsymbol{F} = \lambda^2 \boldsymbol{I} \Rightarrow \boldsymbol{c}^{-1} = \lambda^{-2} \boldsymbol{I} \tag{5.2.23}$$

If we substitute $J = \lambda^3$ and Eq. (5.2.23) into Eq. (5.2.22) then we have:

$$\boldsymbol{s} = K\,(J-1)\lambda \boldsymbol{I} \tag{5.2.24}$$

The Eq. (5.2.24) is a straightforward relation between $\boldsymbol{s}$ and λ. Similar expression can be found between Cauchy stress $\boldsymbol{\sigma}$ and λ if we use the Eq. (3.1.35), that is,

$$\sigma = \frac{1}{J}\boldsymbol{F}\,\boldsymbol{s}\,\boldsymbol{F}^T = \underbrace{\frac{1}{\lambda^3}}_{\frac{1}{J}}\underbrace{\lambda \boldsymbol{I}}_{\boldsymbol{F}}\underbrace{K\,(J-1)\lambda \boldsymbol{I}}_{\boldsymbol{s}}\underbrace{\lambda \boldsymbol{I}}_{\boldsymbol{F}^T} = K\,(J-1)\boldsymbol{I}$$

The above expression can be further clarified as follows:

$$\sigma = K\,(J-1)\boldsymbol{I} = K\frac{dV - dV_0}{dV_0}\boldsymbol{I} = -p\boldsymbol{I}$$

The above expression yields to:

$$p = -K\,(J-1) = -K\frac{dV - dV_0}{dV_0} \tag{5.2.25}$$

The Eq. (5.2.25) is a known expression and been introduced in elasticity and justifies the decomposition technique under the expression $\boldsymbol{F} = J^{1/3}\bar{\boldsymbol{F}}$.

Now we can substitute Eqs. (5.2.5), (5.2.17), (5.2.18), (5.2.6) and (5.2.21) into Eq. (5.2.16) which leads to following expression:

$$s = \underbrace{2C_1 I_3^{-1/3}}_{\frac{\partial w}{\partial I_1}} \underbrace{I}_{\frac{\partial I_1}{\partial \mathbf{c}}} + \underbrace{2C_2 I_3^{-2/3}}_{\frac{\partial w}{\partial I_2}} \underbrace{(I_1 I - c)}_{\frac{\partial I_2}{\partial \mathbf{c}}}$$

$$+ 2\underbrace{\left(-\frac{1}{3} C_1 I_1 I_3^{-4/3} - \frac{2}{3} C_2 I_1 I_3^{-5/3} + \frac{K(J-1)}{2J} \right)}_{\frac{\partial w}{\partial I_3}} \underbrace{I_3 c^{-1}}_{\frac{\partial I_3}{\partial \mathbf{c}}}$$

According to Eq. (5.2.25) we have $\frac{K(J-1)}{2J} I_3 = -p I_3^{1/2}$ therefore the above expression can be abbreviated into this form:

$$s = \left(B_1 I + B_2 c + B_3 c^{-1}\right) - p I_3^{1/2} c^{-1} \tag{5.2.26}$$

In the Eq. (5.2.26) the coefficients B_1, B_2, B_3 are:

$$B_1 = 2C_1 I_3^{-1/3} + 2C_2 I_3^{-2/3} I_1 \qquad B_2 = -2C_2 I_3^{-2/3}$$

$$B_3 = -\frac{2}{3} C_1 I_1 I_3^{-1/3} - \frac{4}{3} C_2 I_1 I_3^{-2/3} \tag{5.2.27}$$

In Eqs. (5.2.26) and (5.2.27) we need to have values for C_1, C_2 and K, for any material these three constants can be determined by experiments and the 2nd Piola stress can be found from Eq. (5.2.26). In a Neo-Hookian materials the expression (5.2.26) can be simplified since in these materials $C_2 = 0$ and $2C_1 = \mu$ then according to Eq. (5.2.27) we have $B_2 = 0$ and $B_1 = \mu I_3^{-1/3}$ also $B_3 = -\frac{\mu}{3} I_1 I_3^{-1/3}$ also considering that $I_3^{-1/3} = J^{-2/3}$ then Eq. (5.2.26) will change to:

$$s = \mu J^{-2/3} \left(I - \frac{1}{3} I_1 c^{-1} \right) - p\, J c^{-1} \tag{5.2.28}$$

In Eq. (5.2.28) we can write $I = c c^{-1}$ then it will change to:

$$s = \mu J^{-2/3} \left(c c^{-1} - \frac{1}{3} I_1 c^{-1} \right) - p\, J c^{-1}$$

The above expression can be written like this:

$$s = \mu J^{-2/3}\left(c - \frac{1}{3}I_1 I\right)c^{-1} - p\,J c^{-1} \tag{5.2.29}$$

According to Eq. (5.2.9) $\bar{c} = J^{-2/3}c$ and Eq. (5.2.11) states that $\bar{I}_1 = I_1 J^{-2/3}$ and if we substitute these in Eq. (5.2.29) then we have:

$$s = \mu\left(\bar{c} - \frac{1}{3}\bar{I}_1 I\right)c^{-1} - p\,J c^{-1} \tag{5.2.30}$$

The deviatory Cauchy Green strain tensor $\mathrm{dev}(\bar{c})$ can be defined in terms of $\bar{c}$ and also $\bar{I}_1$ as follows:

$$\mathrm{dev}(\bar{c}) = \left(\bar{c} - \frac{1}{3}\bar{I}_1 I\right) \tag{5.2.31}$$

The Eq. (5.2.31) is a matrix expression and is similar to the Eq. (4.3.14), which was defined for the stresses. Then Eq. (5.2.31) substituted into Eq. (5.2.30) to change it into following simple form, that is,

$$s = \mu\,\mathrm{dev}(\bar{c})\,c^{-1} - p\,J c^{-1} \tag{5.2.32}$$

Equation (5.2.32) is much better than $\Im_v \tau' = c : \dot{\varepsilon}$ and can be implemented in FEM studies. Equation (5.2.32) unlike Eq. (5.2.25) is not expressed in terms of strain rate tensor $\dot{\varepsilon}$ and a fourth order tensor c, instead the stress tensor s is directly related to the stretch tensor $(\bar{c}, c^{-1})$ which makes Eq. (5.2.32) applicable to FEM studies.

5.3 CONSTITUTIVE EQUATIONS IN PLASTIC REGION

In plastic state what happens is different from elastic state. The material behaves like paste and flows and when the load is removed the material does not return to its initial configuration. The plastic flow situation strongly depends on the yield criteria, which was discussed in Section 4.3. Initially a theory named "Prandtl-Ruess" was developed and then it was formulated by Hill in Cambridge.

He formulated the plastic flow phenomena in terms of an optimization problem. It is stated that subject to a constraint (which is yield criteria)

$f(\sigma)=0$ find a condition (plastic flow) such that power of the plastic deformation $\dot{w}_p=\sigma:\dot{\varepsilon}_p$ becomes maximum, that is,

$$\text{maximise} \quad \dot{w}_p=\sigma:\dot{\varepsilon}_p=\sigma^T\,\dot{\varepsilon}_p \quad \text{if} \quad f(\sigma)=0 \tag{5.3.1a}$$

In Eq. (5.3.1) the σ^T and $\dot{\varepsilon}_p$ are a stress vector (row vector) and plastic strain rate vector (column vector) resulted from the stretching the stress tensor σ and plastic strain rate tensor $\dot{\varepsilon}_p$ and can be displayed similar to Eq. (4.3.19) as follows:

$$\sigma^T=\begin{bmatrix}\sigma_x & \sigma_y & \sigma_z & \tau_{xy} & \tau_{xz} & \tau_{yz}\end{bmatrix} \tag{5.3.1b}$$

$$\dot{\varepsilon}_p^T=\begin{bmatrix}\dot{\varepsilon}_{px} & \dot{\varepsilon}_{py} & \dot{\varepsilon}_{pz} & \dot{\varepsilon}_{pxy} & \dot{\varepsilon}_{pxz} & \dot{\varepsilon}_{pyz}\end{bmatrix} \tag{5.3.1c}$$

We can change Eq. (5.3.1a) to a minimization problem by change of sign in the objective function, that is,

$$\text{minimise} \quad -\dot{w}_p=-\sigma:\dot{\varepsilon}_p=-\sigma^T\,\dot{\varepsilon}_p \quad \text{if} \quad f(\sigma)=0 \tag{5.3.2}$$

In Eqs. (5.3.1a) and (5.3.2) $\dot{w}_p$ is power of the plastic deformation per unit volume. To solve an optimization problem we can use a Lagrange multiplier method. In this method if we want to minimize an objective function, $-\dot{w}_p$ subject to a constraint $f(\sigma)=0$, we need to multiply the constraint by the Lagrange multiplier $\dot{\lambda}$ and add to the objective function, that is,

$$L(\sigma,\dot{\lambda})=-\sigma^T\,\dot{\varepsilon}_p+\dot{\lambda}\,f(\sigma) \tag{5.3.3}$$

In Eq. (5.3.3) the function $L(\sigma,\dot{\lambda})$ is called Lagrangian, for minimizing $L(\sigma,\dot{\lambda})$ it is necessary to take the partial derivative $\frac{\partial L}{\partial \sigma}$ and set it to zero, that is,

$$\frac{\partial L}{\partial \sigma}=0 \Rightarrow -\dot{\varepsilon}_p+\dot{\lambda}\frac{\partial f}{\partial \sigma}=0 \tag{5.3.4}$$

From Eq. (5.3.4) we can conclude that:

$$\dot{\varepsilon}_p=\dot{\lambda}\frac{\partial f}{\partial \sigma} \tag{5.3.5}$$

Since in yield condition $f(\sigma)=0$ we can write:

$$\dot{\lambda}\geq 0 \qquad \dot{\lambda}\, f(\sigma)=0 \tag{5.3.6}$$

The Eqs. (5.3.5) and (5.3.6) in view of mathematicians is called Kuhn-Tucker feasibility condition and it shows that plastic flow strongly depends on the yield criteria. Therefore, it is necessary to examine $f(\sigma)$ which was discussed in Section 4.3 and then find out $\frac{\partial f}{\partial \sigma}$. The yield function was given by Eqs. (4.3.13b) and (4.3.17b) can be written as:

$$f(\sigma)=\sigma_e-\sigma_0=\underbrace{\frac{\sqrt{\begin{array}{l}(\sigma_x-\sigma_y)^2+(\sigma_z-\sigma_y)^2\\ +(\sigma_x-\sigma_z)^2+6\left(\tau_{xy}^2+\tau_{xz}^2+\tau_{zy}^2\right)\end{array}}}{\sqrt{2}}}_{\sigma_e}-\sigma_0=0 \tag{5.3.7}$$

To find out $\frac{\partial f}{\partial \sigma}$ we need to differentiate versus the elements of the vector $\boldsymbol{\sigma}^T$ in Eq. (5.3.1b) we need to use well-known differentiation formula $\frac{d}{dx}\left(\sqrt{u}\right)=\frac{u'}{2\sqrt{u}}$ the vector $\left(\frac{\partial f}{\partial \sigma}\right)^T$ is:

$$\left(\frac{\partial f}{\partial \sigma}\right)^T=\begin{bmatrix}\frac{\partial f}{\partial \sigma_x} & \frac{\partial f}{\partial \sigma_y} & \frac{\partial f}{\partial \sigma_z} & \frac{\partial f}{\partial \tau_{xy}} & \frac{\partial f}{\partial \tau_{xz}} & \frac{\partial f}{\partial \tau_{yz}}\end{bmatrix}$$

Then the elements of $\frac{\partial f}{\partial \sigma}$ are:

$$\frac{\partial f}{\partial \sigma_x}=\frac{1}{\sqrt{2}}\frac{\partial\left((\sigma_x-\sigma_y)^2+(\sigma_x-\sigma_z)^2\right)}{\partial \sigma_x}\frac{1}{2\sqrt{2}\,\sigma_e}=\frac{2\sigma_x-\sigma_y-\sigma_z}{2\sigma_e} \tag{5.3.8a}$$

$$\frac{\partial f}{\partial \sigma_y}=\frac{1}{\sqrt{2}}\frac{\partial\left((\sigma_x-\sigma_y)^2+(\sigma_y-\sigma_z)^2\right)}{\partial \sigma_y}\frac{1}{2\sqrt{2}\,\sigma_e}=\frac{2\sigma_y-\sigma_x-\sigma_z}{2\sigma_e} \tag{5.3.8b}$$

$$\frac{\partial f}{\partial \sigma_z}=\frac{1}{\sqrt{2}}\frac{\partial\left(\left(\sigma_x-\sigma_y\right)^2+\left(\sigma_x-\sigma_z\right)^2\right)}{\partial \sigma_z}\frac{1}{2\sqrt{2}\sigma_e}=\frac{2\sigma_z-\sigma_y-\sigma_x}{2\sigma_e} \tag{5.3.8c}$$

$$\frac{\partial f}{\partial \tau_{xy}}=\frac{1}{\sqrt{2}}\frac{\partial\left(6\tau_{xy}^2\right)}{\partial \tau_{xy}}\frac{1}{2\sqrt{2}\sigma_e}=\frac{6\tau_{xy}}{2\sigma_e} \tag{5.3.8d}$$

$$\frac{\partial f}{\partial \tau_{xz}}=\frac{1}{\sqrt{2}}\frac{\partial\left(6\tau_{xz}^2\right)}{\partial \tau_{xz}}\frac{1}{2\sqrt{2}\sigma_e}=\frac{6\tau_{xz}}{2\sigma_e} \tag{5.3.8e}$$

$$\frac{\partial f}{\partial \tau_{zy}}=\frac{1}{\sqrt{2}}\frac{\partial\left(6\tau_{zy}^2\right)}{\partial \tau_{zy}}\frac{1}{2\sqrt{2}\sigma_e}=\frac{6\tau_{zy}}{2\sigma_e} \tag{5.3.8-f}$$

If we substitute the Eq. (4.3.14) into Eq. (5.3.8), then we can find $\left(\frac{\partial f}{\partial \sigma}\right)^T$ in terms of the elements of stress deviator tensor, that is,

$$\left(\frac{\partial f}{\partial \sigma}\right)^T=\frac{3}{2\sigma_e}\left[\sigma_x' \quad \sigma_y' \quad \sigma_z' \quad 2\tau_{xy} \quad 2\tau_{xz} \quad 2\tau_{yz}\right] \tag{5.3.9}$$

If we set $\left(\frac{\partial f}{\partial \sigma}\right)=\bar{\boldsymbol{a}}$ then Eq. (5.3.9) can be written into this form:

$$\bar{\boldsymbol{a}}=\frac{3}{2\sigma_e}\left(\boldsymbol{L}\bar{\boldsymbol{\sigma}}'\right) \quad \boldsymbol{L}=\begin{bmatrix}1&0&0&0&0&0\\0&1&0&0&0&0\\0&0&1&0&0&0\\0&0&0&2&0&0\\0&0&0&0&2&0\\0&0&0&0&0&2\end{bmatrix} \tag{5.3.10}$$

$$\bar{\boldsymbol{\sigma}}'^T=\left[\sigma_x' \quad \sigma_y' \quad \sigma_z' \quad \tau_{xy} \quad \tau_{xz} \quad \tau_{yz}\right]$$

An alternative form can also be found by using Eqs. (4.3.16) and (4.3.17b) which states that:

$$f(\sigma)=\underbrace{\sqrt{3}\sqrt{\frac{1}{2}\left(\sigma_x'^2+\sigma_y'^2+\sigma_z'^2\right)+\tau_{xy}^2+\tau_{xz}^2+\tau_{zy}^2}}_{\sigma_e}-\sigma_0 \tag{5.3.11}$$

Now differentiating from Eq. (5.3.11) versus components of σ' then we have:

$$\frac{\partial f}{\partial \sigma_x'} = \frac{\sqrt{3}}{2} \frac{\partial \sigma_x'^2}{\partial \sigma_x'} \frac{\sqrt{3}}{2\,\sigma_e} = \frac{3}{2\sigma_e} \sigma_x' \tag{5.3.12a}$$

$$\frac{\partial f}{\partial \sigma_y'} = \frac{\sqrt{3}}{2} \frac{\partial \sigma_y'^2}{\partial \sigma_y'} \frac{\sqrt{3}}{2\,\sigma_e} = \frac{3}{2\sigma_e} \sigma_y' \tag{5.3.12b}$$

$$\frac{\partial f}{\partial \sigma_z'} = \frac{\sqrt{3}}{2} \frac{\partial \sigma_z'^2}{\partial \sigma_z'} \frac{\sqrt{3}}{2\,\sigma_e} = \frac{3}{2\sigma_e} \sigma_z' \tag{5.3.12c}$$

The Eqs. (5.3.8d), (5.3.8e) and (5.3.8f) remain the same, the above Eq. (5.3.12) together with Eqs. (5.3.8d), (5.3.8e) and (5.3.8f) can be written into this form:

$$\left(\frac{\partial f}{\partial \bar{\sigma}'}\right) = \bar{\boldsymbol{a}} = \left(\frac{\partial f}{\partial \bar{\sigma}}\right) = \frac{3}{2\sigma_e}\left(\boldsymbol{L}\bar{\sigma}'\right) \tag{5.3.12}$$

In Eqs. (5.3.10) and (5.3.12) $\bar{\boldsymbol{a}}$, $\bar{\sigma}$ and $\bar{\sigma}'$ are column vectors (with six elements) that contain elements of the matrices $\boldsymbol{a}$, $\bar{\sigma}$ and σ' (defined by Eq. (4.3.18b)). Obviously the Eq. (5.3.12) can be expressed in terms of $\boldsymbol{a}$, σ and σ' as well (without appearance of $\boldsymbol{L}$) into this form:

$$\left(\frac{\partial f}{\partial \sigma'}\right) = \boldsymbol{a} = \left(\frac{\partial f}{\partial \sigma}\right) = \frac{3}{2\sigma_e}\sigma' \tag{5.3.13}$$

Now we need to clarify the physical meaning of the Lagrange multipliers $\dot{\lambda}$. In mechanics these multipliers are the internal forces. To answer to the question we can rewrite Eq. (5.3.5) into the form $\dot{\boldsymbol{\varepsilon}}_p = \dot{\lambda}\,\boldsymbol{a}$ remembering that $\dot{\boldsymbol{\varepsilon}}_p$ is the plastic strain rate tensor. Similar to Eq. (5.3.16) by which σ_e was defined, herein we define equivalent plastic strain rate tensor $\dot{\varepsilon}_{ps}$ as follows:

$$\dot{\varepsilon}_{ps} = \sqrt{\frac{2}{3}}\sqrt{\dot{\varepsilon}_{px}^2 + \dot{\varepsilon}_{py}^2 + \dot{\varepsilon}_{py}^2 + \frac{1}{2}\left(\dot{\gamma}_{xy}^2 + \dot{\gamma}_{xz}^2 + \dot{\gamma}_{yz}^2\right)} \tag{5.3.14}$$

The definition in Eq. (5.3.14) includes all the elements of the plastic strain rate tensor $\dot{\boldsymbol{\varepsilon}}_p$, that is,

$$\dot{\boldsymbol{\varepsilon}}_p = \begin{bmatrix} \dot{\varepsilon}_{px} & \dot{\gamma}_{xy}/2 & \dot{\gamma}_{xz}/2 \\ \dot{\gamma}_{yx}/2 & \dot{\varepsilon}_{py} & \dot{\gamma}_{yz}/2 \\ \dot{\gamma}_{zx}/2 & \dot{\gamma}_{zy}/2 & \dot{\varepsilon}_{pz} \end{bmatrix} \quad (5.3.15)$$

Similar to Eq. (5.3.15) the strain tensor (not rate) was also defined in previous chapters. According to the scalar product of the matrices defined in Section 4.2, we can express $\dot{\varepsilon}_{ps}$ (scalar quantity) in terms of the elements of the matrix $\dot{\boldsymbol{\varepsilon}}_p$ into this form:

$$\dot{\varepsilon}_{ps} = \sqrt{\frac{2}{3}}\sqrt{\dot{\boldsymbol{\varepsilon}}_p : \dot{\boldsymbol{\varepsilon}}_p} \quad (5.3.16)$$

If we substitute Eq. (5.3.13) into $\dot{\boldsymbol{\varepsilon}}_p = \dot{\lambda}\,\boldsymbol{a}$ then we have $\dot{\boldsymbol{\varepsilon}}_p = \dfrac{3\dot{\lambda}}{2\sigma_e}\sigma'$ then we substitute into Eq. (5.3.16) and we have:

$$\dot{\varepsilon}_{ps} = \sqrt{\frac{2}{3}}\sqrt{\left(\frac{3\dot{\lambda}}{2\sigma_e}\sigma'\right):\left(\frac{3\dot{\lambda}}{2\sigma_e}\sigma'\right)} \quad (5.3.17)$$

The Eq. (5.3.17) can be simplified into:

$$\dot{\varepsilon}_{ps} = \sqrt{\frac{3}{2}}\frac{\dot{\lambda}}{\sigma_e}\sqrt{(\sigma'):(\sigma')} \quad (5.3.18)$$

According to the Eqs. (4.3.17b) and (4.3.18a) it is obvious that $\sqrt{\frac{3}{2}}\sqrt{(\sigma'):(\sigma')} = \sigma_e$ and if we substitute this into Eq. (5.3.19) then we have:

$$\dot{\lambda} = \dot{\varepsilon}_{ps} \quad (5.3.19)$$

Equation (5.3.19) provides a physical interpretation for the Lagrange multiplier $\dot{\lambda}$, saying that it is identical to equivalent plastic strain rate $\dot{\varepsilon}_{ps}$ which is defined by Eqs. (5.3.14), (5.3.15) and (5.3.16). Although we are familiar with strain tensors in small deformation, we can have sense of strain rate tensor $\dot{\varepsilon}_p$ in small deformation. In large deformation though, clarification of $\dot{\varepsilon}_p$ is a formidable task and different views exists one can see the details of the controversy in Simo [4], in the rest of this section we are trying to explain what is exactly $\dot{\varepsilon}_p$ tensor in large deformations?

According to Fig. 5.2, the whole deformation expressed by tensor $\boldsymbol{F}$ can be decomposed into plastic and elastic parts. We can clarify later why plastic part comes first? An element in initial configuration $d\boldsymbol{X}$ after the plastic deformation expressed by tensor $\boldsymbol{F}_p$ will change to $d\hat{\boldsymbol{x}}$ under this transformation:

$$d\hat{\boldsymbol{x}} = \frac{\partial \hat{\boldsymbol{x}}}{\partial \boldsymbol{X}} d\boldsymbol{X} = \boldsymbol{F}_p \, d\boldsymbol{X} \tag{5.3.20}$$

For the elastic part expressed by $\boldsymbol{F}_e$ the element $d\hat{\boldsymbol{x}}$ will change to $d\boldsymbol{x}$ under this transformation:

$$d\boldsymbol{x} = \frac{\partial \boldsymbol{x}}{\partial \hat{\boldsymbol{x}}} d\hat{\boldsymbol{x}} = \boldsymbol{F}_e \, d\hat{\boldsymbol{x}} \tag{5.3.21}$$

We substitute $d\hat{\boldsymbol{x}}$ from Eqs. (5.3.20) into (5.3.21) then we have:

$$d\boldsymbol{x} = \frac{\partial \boldsymbol{x}}{\partial \boldsymbol{X}} d\boldsymbol{X} = \boldsymbol{F} \, d\boldsymbol{X} = \boldsymbol{F}_e \, \boldsymbol{F}_p \, d\boldsymbol{X} \tag{5.3.22}$$

Such that from Eq. (5.3.22) we can find decomposition formula, that is,

$$\boldsymbol{F} = \boldsymbol{F}_e \, \boldsymbol{F}_p \tag{5.3.23}$$

In order to clarify the $\dot{\varepsilon}_p$ concept we rewrite definition (4.2.15) as follows:

$$\boldsymbol{L} = \frac{\partial \boldsymbol{v}}{\partial \boldsymbol{x}} = \dot{\boldsymbol{F}} \, \boldsymbol{F}^{-1} \tag{5.3.24}$$

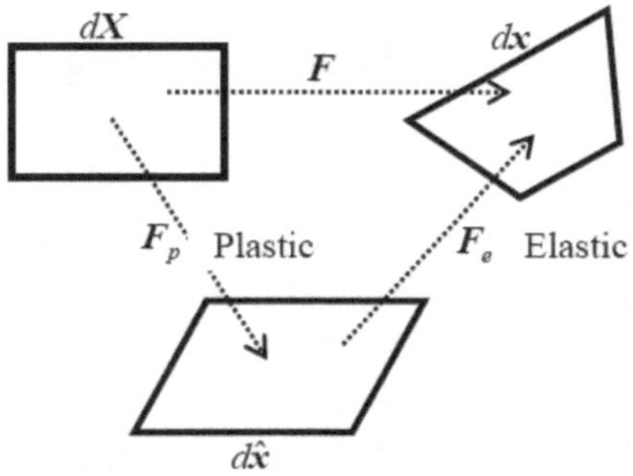

FIGURE 5.2 Decomposition into elastic and plastic deformation.

Now we substitute Eq. (5.3.23) into Eq. (5.3.24) then we have:

$$L = \frac{d\left(F_e F_p\right)}{dt}\left(F_e F_p\right)^{-1} = \left(\dot{F}_e F_p + F_e \dot{F}_p\right) F_p^{-1} F_e^{-1}$$

The above expression can be simplified to:

$$L = \dot{F}_e F_e^{-1} + F_e \dot{F}_p F_p^{-1} F_e^{-1} \tag{5.3.25}$$

According to Eq. (5.3.25) we define elastic and plastic velocity gradient tensor as follows:

$$l_e = \dot{F}_e F_e^{-1} \qquad l_p = F_e \dot{F}_p F_p^{-1} F_e^{-1} \tag{5.3.26a}$$

From Eq. (5.3.25) and definition (5.3.26a) we find that:

$$L = l_e + l_p \tag{5.3.26b}$$

Equation (5.3.26b) explains that the velocity gradient L can splits into two parts, the elastic part l_e contains elastic deformation $\dot{F}_e F_e^{-1}$, whereas the plastic part l_p contains both plastic and elastic deformation $F_e \dot{F}_p F_p^{-1} F_e^{-1}$. According to Eq. (5.3.23) deformation gradient is multiplicative type of decomposition, whereas velocity gradient in Eq. (5.3.26b) is additive type of decomposition. The latter in Eq. (5.3.26b) coincides with the classical plasticity theory therefore we can justify further discussion about it. If we define L_p in the intermediate coordinate $\hat{x}$ similar to the definition in Eq. (4.2.15), then we have:

$$L_p = \dot{F}_p F_p^{-1} \tag{5.3.27}$$

The Eq. (5.3.27) represents the plastic deformation only in which $dX \Rightarrow d\hat{x}$ such that the elastic deformation is ignored. If we substitute Eq. (5.3.27) into Eq. (5.3.26a) then we have:

$$l_p = F_e L_p F_e^{-1} \tag{5.3.28}$$

In Eq. (5.3.28) the l_p is a tensor based on the deformed coordinate x but L_p is also tensor based on the intermediate coordinate $\hat{x}$, they can be transformed according to $l_p = F_e L_p F_e^{-1}$ which can be justified only if we assume $\boldsymbol{F}_e^T$ and $\boldsymbol{F}_e^{-1}$ are identical. Therefore, we can say Eq. (5.3.28) can be justified only because Eq. (5.3.26b) is a valid expression. Now we can define plastic strain tensor $\dot{\boldsymbol{\varepsilon}}_p$ based on the definition in Eq. (4.2.6), that is,

$$\dot{\boldsymbol{\varepsilon}}_p = \frac{1}{2}\left(\boldsymbol{l}_p + \boldsymbol{l}_p^T\right) \quad \dot{\boldsymbol{E}}_p = \frac{1}{2}\left(\boldsymbol{L}_p + \boldsymbol{L}_p^T\right)$$

The tensor $\dot{\boldsymbol{\varepsilon}}_p$ is defined on deformed frame $\boldsymbol{x}$, and the tensor $\dot{\boldsymbol{E}}_p$ is defined on the intermediate frame $\hat{\boldsymbol{x}}$ we can also implement transformation rule (5.3.28) to these, that is,

$$\dot{\boldsymbol{\varepsilon}}_{\mathrm{p}} = \boldsymbol{F}_e \dot{\boldsymbol{E}}_p \boldsymbol{F}_e^{-1} = \frac{1}{2}\boldsymbol{F}_e\left(\boldsymbol{L}_p + \boldsymbol{L}_p^T\right)\boldsymbol{F}_e^{-1} \tag{5.3.29}$$

We can accept Eq. (5.3.29) for $\dot{\boldsymbol{\varepsilon}}_p$ based on the validity of Eq. (5.3.26). Assuming there is not any reservation on Eq. (5.3.26) we can substitute Eq. (5.3.27) into Eq. (5.3.29) to expand it further, that is,

$$\dot{\boldsymbol{\varepsilon}}_p = \frac{1}{2}\boldsymbol{F}_{\mathrm{e}}\left(\dot{\boldsymbol{F}}_p\boldsymbol{F}_p^{-1} + \boldsymbol{F}_p^{-T}\dot{\boldsymbol{F}}_p^T\right)\boldsymbol{F}_e^{-1} \tag{5.3.30}$$

Equation (5.3.30) provides full information about $\dot{\boldsymbol{\varepsilon}}_p$ if the elastic and plastic deformation gradient are known. Now we can define plastic right Cauchy Green tensor like this:

$$\boldsymbol{c}_p = \boldsymbol{F}_p^T\boldsymbol{F}_p \tag{5.3.31}$$

Also elastic left Cauchy Green tensor by:

$$\boldsymbol{b}_e = \boldsymbol{F}_e\boldsymbol{F}_e^T \tag{5.3.32}$$

According to Eq. (5.3.23) we can write:

$$\boldsymbol{F}_e = \boldsymbol{F}\boldsymbol{F}_p^{-1} \Rightarrow \boldsymbol{F}_e^T = \boldsymbol{F}_p^{-T}\boldsymbol{F}^T \tag{5.3.33}$$

If we substitute Eq. (5.3.33) into Eq. (5.3.32) then we have:

$$\boldsymbol{b}_e = \boldsymbol{F}\boldsymbol{F}_p^{-1}\boldsymbol{F}_p^{-T}\boldsymbol{F}^T = \boldsymbol{F}\left(\boldsymbol{F}_p^T\boldsymbol{F}_p\right)^{-1}\boldsymbol{F}^T$$

From Eq. (5.3.31) we have $\boldsymbol{c}_p = \boldsymbol{F}_p^T \boldsymbol{F}_p$ if we substitute this into above expression we have:

$$\boldsymbol{b}_e = \boldsymbol{F}\,\boldsymbol{c}_p^{-1}\boldsymbol{F}^T \tag{5.3.34}$$

Equation (5.3.34) shows that the tensor $\boldsymbol{c}_p^{-1}$ that represents plastic deformation, is related to $\boldsymbol{b}_e$ which represents elastic deformation via the "pull back" and "push forward" operators as described in Section 5.1, like this:

$$\underbrace{\boldsymbol{c}_p^{-1}}_{dx} \underset{\text{pull back}}{\underbrace{\Rightarrow}} \underbrace{\boldsymbol{b}_e}_{dX} \qquad \underbrace{\boldsymbol{b}_e}_{dX} \underset{\text{push forward}}{\underbrace{\Rightarrow}} \underbrace{\boldsymbol{c}_p^{-1}}_{dx} \tag{5.3.35}$$

The other version of the Eq. (5.3.34) will be obtained by a pre and post multiplying to $\boldsymbol{F}^{-1}$ and $\boldsymbol{F}^{-T}$, then we have:

$$\boldsymbol{c}_p^{-1} = \boldsymbol{F}^{-1}\boldsymbol{b}_e\,\boldsymbol{F}^{-T} \tag{5.3.36}$$

We studied the transformations like Eqs. (5.3.35) and (5.3.36) in Section 5.1, where in Eq. (5.1.10) we defined Lie derivative for the tensor quantities between two frames. If we implement the Lie derivative rule (5.1.11a) on the Eq. (5.3.36) then we have:

$$\Im_v \boldsymbol{b}_e = \boldsymbol{F}\left(\frac{D}{Dt}\left(\underbrace{\boldsymbol{F}^{-1}\,\boldsymbol{b}_e\,\boldsymbol{F}^{-T}}_{\text{pull back}}\right)\right)\boldsymbol{F}^T$$

If we substitute Eq. (5.3.36) into the above expression then we have:

$$\Im_v \boldsymbol{b}_e = \boldsymbol{F}\,\dot{\boldsymbol{c}}_p^{-1}\boldsymbol{F}^T \tag{5.3.37}$$

Equation (5.3.37) explains that the Lie derivative concept can describe the relationship between elastic and plastic deformation. Now we invert the Eq. (5.3.34) and then we have:

$$\boldsymbol{b}_e^{-1} = \boldsymbol{F}^{-T}\boldsymbol{c}_p\,\boldsymbol{F}^{-1} \tag{5.3.38}$$

If we multiply Eqs. (5.3.37) to (5.3.38) we have:

$$-\frac{1}{2}(\Im_v \boldsymbol{b}_e)\boldsymbol{b}_e^{-1} = -\frac{1}{2}\left(\boldsymbol{F}\,\dot{\boldsymbol{c}}_p^{-1}\boldsymbol{F}^T\right)\left(\boldsymbol{F}^{-T}\boldsymbol{c}_p\,\boldsymbol{F}^{-1}\right) = -\frac{1}{2}\boldsymbol{F}\,\dot{\boldsymbol{c}}_p^{-1}\boldsymbol{c}_p\,\boldsymbol{F}^{-1} \tag{5.3.39}$$

This can be followed by taking material time derivative of the expression, $\boldsymbol{c}_p^{-1}\boldsymbol{c}_p = \boldsymbol{I}$ then we have:

$$\dot{\boldsymbol{c}}_p^{-1}\boldsymbol{c}_p + \boldsymbol{c}_p^{-1}\dot{\boldsymbol{c}}_p = 0 \Rightarrow \dot{\boldsymbol{c}}_p^{-1}\boldsymbol{c}_p = -\boldsymbol{c}_p^{-1}\dot{\boldsymbol{c}}_p \tag{5.3.40a}$$

From $\boldsymbol{F} = \boldsymbol{F}_e\,\boldsymbol{F}_p$ we have $\boldsymbol{F}^{-1} = \boldsymbol{F}_p^{-1}\,\boldsymbol{F}_e^{-1}$ and if we substitute these, together with Eq. (5.3.40) into Eq. (5.3.39) then we have:

$$-\frac{1}{2}\left(\Im_v\,\boldsymbol{b}_e\right)\boldsymbol{b}_e^{-1} = \frac{1}{2}\boldsymbol{F}_e\,\boldsymbol{F}_p\,\boldsymbol{c}_p^{-1}\dot{\boldsymbol{c}}_p\,\boldsymbol{F}_p^{-1}\,\boldsymbol{F}_e^{-1} \tag{5.3.40b}$$

Exercise 5.3.1: *Simplify the* Eq. (5.3.40b)

Solution: From Eq. (5.3.31) we have $\boldsymbol{c}_p = \boldsymbol{F}_p^T\boldsymbol{F}_p$ therefore we have $\dot{\boldsymbol{c}}_p = \dot{\boldsymbol{F}}_p^T\boldsymbol{F}_p + \boldsymbol{F}_p^T\dot{\boldsymbol{F}}_p$ and also $\boldsymbol{c}_p^{-1} = \boldsymbol{F}_p^{-1}\boldsymbol{F}_p^{-T}$, then we substitute these into Eq. (5.3.40b) and we have:

$$-\frac{1}{2}\left(\Im_v\,\boldsymbol{b}_e\right)\boldsymbol{b}_e^{-1} = \frac{1}{2}\boldsymbol{F}_e\,\boldsymbol{F}_p\left(\boldsymbol{F}_p^{-1}\boldsymbol{F}_p^{-T}\right)\left(\dot{\boldsymbol{F}}_p^T\boldsymbol{F}_p + \boldsymbol{F}_p^T\dot{\boldsymbol{F}}_p\right)\boldsymbol{F}_p^{-1}\,\boldsymbol{F}_e^{-1}$$

The above expression can be expanded after multiplication into:

$$\begin{aligned}-\frac{1}{2}\left(\Im_v\,\boldsymbol{b}_e\right)\boldsymbol{b}_e^{-1} &= \frac{1}{2}\boldsymbol{F}_e\,\underbrace{\boldsymbol{F}_p\boldsymbol{F}_p^{-1}}_{\boldsymbol{I}}\,\boldsymbol{F}_p^{-T}\dot{\boldsymbol{F}}_p^T\,\underbrace{\boldsymbol{F}_p\,\boldsymbol{F}_p^{-1}}_{\boldsymbol{I}}\,\boldsymbol{F}_e^{-1}\\ &+\frac{1}{2}\boldsymbol{F}_e\,\underbrace{\boldsymbol{F}_p\boldsymbol{F}_p^{-1}\boldsymbol{F}_p^{-T}\boldsymbol{F}_p^T}_{\boldsymbol{I}}\,\dot{\boldsymbol{F}}_p\,\boldsymbol{F}_p^{-1}\,\boldsymbol{F}_e^{-}\end{aligned}$$

The above expression easily yields to:

$$-\frac{1}{2}\left(\Im_v\,\boldsymbol{b}_e\right)\boldsymbol{b}_e^{-1} = \frac{1}{2}\boldsymbol{F}_e\left(\boldsymbol{F}_p^{-T}\dot{\boldsymbol{F}}_p^T + \dot{\boldsymbol{F}}_p\,\boldsymbol{F}_p^{-1}\right)\boldsymbol{F}_e^{-1} \tag{5.3.41}$$

If we substitute Eq. (5.3.30) into Eq. (5.3.41), that is, comparing their right sides, then we have:

$$\dot{\boldsymbol{\varepsilon}}_p = -\frac{1}{2}\left(\Im_v\,\boldsymbol{b}_e\right)\boldsymbol{b}_e^{-1} \tag{5.3.42}$$

Equation (5.3.42) is the simplest form of the strain rate tensor $\dot{\varepsilon}_p$ in large elastic-plastic deformation, which is also valid. If we compare Eqs. (5.3.42) and (5.3.5) then we have:

$$\dot{\varepsilon}_p = -\frac{1}{2}\left(\Im_v \boldsymbol{b}_e\right)\boldsymbol{b}_e^{-1} = \dot{\lambda}\frac{\partial f}{\partial \sigma} \tag{5.3.43}$$

From Eq. (5.3.19) we substitute for $\dot{\lambda}$, and from Eq. (5.3.13) we substitute for $\frac{\partial f}{\partial \sigma}$ into Eq. (5.3.43), then we have:

$$\left(\Im_v \boldsymbol{b}_e\right)\boldsymbol{b}_e^{-1} = -3\dot{\varepsilon}_{ps}\frac{\sigma'}{\sigma_e} \tag{5.3.44}$$

The Eq. (5.3.44) is the simplest and accurate relationship between deviatory Cauchy stress tensor $\boldsymbol{\sigma}'$ and elastic part of deformation $\boldsymbol{b}_e$ and also equivalent plastic strain rate $\dot{\varepsilon}_{ps}$. However, for large elastic deformation the relationship is described in terms of $\boldsymbol{s}$ or 2nd Piola Kirchhoff stress given by Eq. (5.2.32).

Equation (5.3.44) is based on constant yield stress σ_0 in Eq. (5.3.11), in this case the material is called elastic perfectly plastic (e-p-p). In elastic plastic strain hardening materials the Eq. (5.3.11) will change such that yield surface is:

$$f(\sigma) = \sigma_e - \sigma_0\left(\varepsilon_{ps}\right) = 0 \tag{5.3.45a}$$

The equivalent plastic strain ε_{ps} in Eq. (5.3.45a) can be found by time integration of $\dot{\varepsilon}_{ps}$, that is,

$$\varepsilon_{ps} = \int_0^t \dot{\varepsilon}_{ps}\, dt \tag{5.3.45b}$$

Considering strain hardening obviously changes Eq. (5.3.44) but similar expressions to Eq. (5.3.44) can be derived if the strain stress curves can be obtained by the experiments. This is beyond the scope of this manuscript.

There is a special case, which is called isotropic strain hardening, in which $\sigma_0(\varepsilon_{ps}) = \sigma_0 + K\varepsilon_{ps}$. Therefore, Eq. (5.3.45a) will change to:

$$f(\sigma) = \sigma_e - (\sigma_0 + K\varepsilon_{ps}) = 0 \tag{5.3.46}$$

Equations (5.3.44) in strain hardening case, Eqs. (5.3.45) and (5.3.46) are system of nonlinear equations that cannot be solved analytically but they may be solved numerically. However, Eq. (5.3.44) for e-p-p materials together with Eq. (5.2.32) describes the material behavior in plastic and elastic parts, respectively. Also they cannot be solved analytically except in very simple cases. The thorough determination of the stresses and deformation these days are easier by numerical methods including FEM studies.

KEYWORDS

- **Cauchy-Green tensor**
- **Green strain tensor**
- **Hook tensor**
- **pull back operator**
- **push forward operator**
- **Lie derivative**

PART 2

MECHANICS OF FRACTURE

CHAPTER 6

APPLICATION OF COMPLEX VARIABLE METHOD IN LINEAR ELASTICITY

CONTENTS

6.1 REVIEW OF ELASTICITY THEORY

It should be remembered that small strains in elasticity are:

$$\varepsilon_x = \frac{\partial u}{\partial x} \qquad \varepsilon_y = \frac{\partial v}{\partial x} \qquad \text{and} \qquad \varepsilon_{xy} = \frac{1}{2}\left(\frac{\partial u}{\partial y} + \frac{\partial v}{\partial x}\right) \tag{6.1.1}$$

In (6.1.1), u and v are displacement in x and y directions and ε_x, ε_y and ε_{xy} are the strains resulted by the stresses σ according to the Hook's law as follows:

$$\varepsilon_x = \frac{1}{E}\left[\sigma_x - \upsilon\left(\sigma_y + \sigma_z\right)\right] \quad \varepsilon_y = \frac{1}{E}\left[\sigma_y - \upsilon\left(\sigma_x + \sigma_z\right)\right]$$

$$\varepsilon_z = \frac{1}{E}\left[\sigma_z - \upsilon\left(\sigma_y + \sigma_x\right)\right] \tag{6.1.2}$$

In two-dimensional problems, plane stress types are defined by:

$$\sigma_z = \tau_{zx} = \tau_{zy} = 0 \tag{6.1.3}$$

In Eq. (6.1.3) τ is shear stress, which is related to the twist or ε_{xy}, as follows:

$$\gamma_{xy} = \frac{\tau_{xy}}{G} \quad \gamma_{zy} = \frac{\tau_{zy}}{G} \quad \gamma_{zx} = \frac{\tau_{zx}}{G} \tag{6.1.4}$$

The γ in Eq. (6.1.4) is twice the twist, and for isotropic material, the modulus of elasticity E is related to shear modulus G via the Poisson ratio υ as follows:

$$\varepsilon_{xy} = \frac{\gamma_{xy}}{2} \quad \varepsilon_{zy} = \frac{\gamma_{zy}}{2} \quad \varepsilon_{zx} = \frac{\gamma_{zx}}{2} \tag{6.1.5}$$

$$G = \frac{E}{2(1+\upsilon)} \tag{6.1.6}$$

By substituting equations (6.1.5) and (6.1.6) into (6.1.4) we have:

$$\varepsilon_{xy} = \frac{1+\upsilon}{E}\tau_{xy} \quad \varepsilon_{zy} = \frac{1+\upsilon}{E}\tau_{zy} \quad \varepsilon_{xz} = \frac{1+\upsilon}{E}\tau_{xz} \tag{6.1.7}$$

Although in plain stress situation $\sigma_z = \tau_{zx} = \tau_{zy} = 0$, but the strain $\varepsilon_z \neq 0$. For Mode III fracture the governing stress state, will change and we discuss those later. For plane stress status the Eq. (6.1.2) will be simplified into:

$$\varepsilon_x = \frac{1}{E}\left[\sigma_x - \upsilon\sigma_y\right] \tag{6.1.8}$$

$$\varepsilon_y = \frac{1}{E}\left[\sigma_y - \upsilon\sigma_x\right] \tag{6.1.9}$$

$$\varepsilon_{xy} = \frac{1+\upsilon}{E}\tau_{xy} \tag{6.1.10}$$

In two-dimensional problems the equilibrium equation will be simplified to:

$$\frac{\partial \sigma_x}{\partial x} + \frac{\partial \tau_{xy}}{\partial y} = 0 \tag{6.1.11}$$

$$\frac{\partial \sigma_y}{\partial y} + \frac{\partial \tau_{yx}}{\partial x} = 0 \tag{6.1.12}$$

By differentiating from Eq. (6.1.1) we can write:

$$\frac{\partial^2 \varepsilon_x}{\partial y^2} = \frac{\partial^2}{\partial y^2}\left(\frac{\partial u}{\partial x}\right) = \frac{\partial^3 u}{\partial x\,\partial y^2} \tag{6.1.13}$$

$$\frac{\partial^2 \varepsilon_y}{\partial x^2} = \frac{\partial^2}{\partial x^2}\left(\frac{\partial v}{\partial y}\right) = \frac{\partial^3 v}{\partial y\,\partial x^2} \tag{6.1.14}$$

$2\frac{\partial^2 \varepsilon_{xy}}{\partial x\,\partial y} = 2\frac{1}{2}\frac{\partial^2}{\partial x\,\partial y}\left(\frac{\partial v}{\partial x} + \frac{\partial u}{\partial y}\right)$ and after simplification:

$$2\frac{\partial^2 \varepsilon_{xy}}{\partial x\,\partial y} = \frac{\partial^3 v}{\partial x^2\,\partial y} + \frac{\partial^3 u}{\partial x\,\partial y^2} \tag{6.1.15}$$

It is obvious that equations (6.1.13), (6.1.14) and (6.1.15) can be combined together by:

$$\frac{\partial^2 \varepsilon_x}{\partial y^2} + \frac{\partial^2 \varepsilon_y}{\partial x^2} - 2\frac{\partial^2 \varepsilon_{xy}}{\partial x\,\partial y} = 0 \tag{6.1.16}$$

The Eq. (6.1.16) is called compatibility relationship and each term of it can be expanded by using the Eqs. (6.1.8), (6.1.9) and (6.1.10) as follows:

$$\frac{\partial^2 \varepsilon_x}{\partial y^2} = \frac{1}{E}\left(\frac{\partial^2 \sigma_x}{\partial y^2} - \upsilon\frac{\partial^2 \sigma_y}{\partial y^2}\right) \tag{6.1.17}$$

$$\frac{\partial^2 \varepsilon_y}{\partial x^2} = \frac{1}{E}\left(\frac{\partial^2 \sigma_y}{\partial x^2} - \upsilon \frac{\partial^2 \sigma_x}{\partial x^2}\right) \tag{6.1.18}$$

$$2\frac{\partial^2 \varepsilon_{xy}}{\partial x\,\partial y} = \frac{2(1+\upsilon)}{E}\frac{\partial^2 \tau_{xy}}{\partial x\,\partial y} \tag{6.1.19}$$

If we substitute Eqs. (6.1.17), (6.1.18) and (6.1.19) into Eq. (6.1.16) we have:

$$\frac{1}{E}\left(\frac{\partial^2 \sigma_x}{\partial y^2} - \upsilon \frac{\partial^2 \sigma_y}{\partial y^2}\right) + \frac{1}{E}\left(\frac{\partial^2 \sigma_y}{\partial x^2} - \upsilon \frac{\partial^2 \sigma_x}{\partial x^2}\right) - \frac{2(1+\upsilon)}{E}\frac{\partial^2 \tau_{xy}}{\partial x\,\partial y} = 0 \tag{6.1.20}$$

Now we define the stress function F so that it can express the stresses as:

$$\sigma_x = \frac{\partial^2 F}{\partial y^2} \qquad \sigma_y = \frac{\partial^2 F}{\partial x^2} \qquad \tau_{xy} = -\frac{\partial^2 F}{\partial y\,\partial x} \tag{6.1.21}$$

The above forms in Eq. (6.1.21) are capable of satisfying the equilibrium equations as can be seen below:

$$\frac{\partial}{\partial x}\left(\frac{\partial^2 F}{\partial y^2}\right) + \frac{\partial}{\partial y}\left(-\frac{\partial^2 F}{\partial x\,\partial y}\right) = 0 \quad \text{and} \quad \frac{\partial}{\partial y}\left(\frac{\partial^2 F}{\partial x^2}\right) + \frac{\partial}{\partial x}\left(-\frac{\partial^2 F}{\partial x\,\partial y}\right) = 0$$

Now, we substitute Eq. (6.1.20) into Eq. (6.1.21) and write down all the terms:

$$\frac{1}{E}\left(\frac{\partial^2}{\partial y^2}\left(\frac{\partial^2 F}{\partial y^2}\right) - \upsilon \frac{\partial^2}{\partial y^2}\left(\frac{\partial^2 F}{\partial x^2}\right)\right) + \frac{1}{E}\left(\frac{\partial^2}{\partial x^2}\left(\frac{\partial^2 F}{\partial x^2}\right) - \upsilon \frac{\partial^2}{\partial x^2}\left(\frac{\partial^2 F}{\partial y^2}\right)\right)$$
$$-\frac{2(1+\upsilon)}{E}\frac{\partial^2}{\partial x\,\partial y}\left(-\frac{\partial^2 F}{\partial y\,\partial x}\right) = 0$$

It is obvious that the above form can be simplified to:

$$\frac{\partial^4 F}{\partial y^4} + \frac{\partial^4 F}{\partial x^4} + 2\frac{\partial^4 F}{\partial y^2\,\partial x^2} = 0 \text{ or } \left(\frac{\partial^2}{\partial y^2} + \frac{\partial^2}{\partial x^2}\right)\left(\frac{\partial^2 F}{\partial y^2} + \frac{\partial^2 F}{\partial x^2}\right) = 0 \text{ i.e.}$$

$$\nabla^2\left(\nabla^2(F)\right) = 0 \tag{6.1.22}$$

The Eq. (6.1.22) is called bi-harmonic equation and through solving it the stress function can be determined. These solutions will be followed in the incoming parts, since they are 6tal in determination of stresses around the crack tips.

F is also called Airy stress function, herein we prove that this function is also acceptable for two dimensional problems with plane strain state. In plane strain state we have:

$$\varepsilon_z = \varepsilon_{zx} = \varepsilon_{zy} = 0 \tag{6.1.23}$$

If we examine ε_z from Eq. (6.1.2) then we have:

$$\varepsilon_z = \frac{1}{E}\left[\sigma_z - \upsilon\left(\sigma_y + \sigma_x\right)\right] = 0 \Rightarrow \sigma_z = \left(\sigma_y + \sigma_x\right)\upsilon \tag{6.1.24}$$

Exercise 6.1.1: Prove that bi-harmonic equation is valid for plane strain case.

Solution: Substitute Eq. (6.1.24) into Eq. (6.1.2) for ε_x and ε_y, then we have:

$$\varepsilon_x = \frac{1}{E}\left[\sigma_x - \upsilon\sigma_y - \upsilon^2\left(\sigma_y + \sigma_x\right)\right] = \frac{1}{E}\left[\left(1-\upsilon^2\right)\sigma_x - \upsilon\left(1+\upsilon\right)\sigma_y\right] \tag{6.1.25}$$

$$\varepsilon_y = \frac{1}{E}\left[\sigma_y - \upsilon\sigma_x - \upsilon^2\left(\sigma_y + \sigma_x\right)\right] = \frac{1}{E}\left[\left(1-\upsilon^2\right)\sigma_y - \upsilon\left(1+\upsilon\right)\sigma_x\right] \tag{6.1.26}$$

Since, the compatibility Eq. (6.1.16) is also valid for plane strain case and so the Eq. (6.1.10), we can substitute Eqs. (6.1.25), (6.1.26) and (6.1.10) into Eq. (6.1.16) and then we have:

$$\frac{1}{E}\left(\left(1-\upsilon^2\right)\frac{\partial^2\sigma_x}{\partial y^2} - \upsilon\left(1+\upsilon\right)\frac{\partial^2\sigma_y}{\partial y^2}\right) + \frac{1}{E}\left(\left(1-\upsilon^2\right)\frac{\partial^2\sigma_y}{\partial x^2} - \upsilon\left(1+\upsilon\right)\frac{\partial^2\sigma_x}{\partial x^2}\right) - \frac{2\left(1+\upsilon\right)}{E}\frac{\partial^2\tau_{xy}}{\partial x\,\partial y} = 0 \tag{6.1.27}$$

The equilibrium Eqs. (6.1.11) and (6.1.12) are also valid for plane strain case, and therefore the stress functions (6.1.21) can be substituted into Eq. (6.1.27) and the result is a long expression,

$$\frac{1}{E}\left(\left(1-\upsilon^2\right)\frac{\partial^2}{\partial y^2}\left(\frac{\partial^2 F}{\partial y^2}\right)-\upsilon\left(1+\upsilon\right)\frac{\partial^2}{\partial y^2}\left(\frac{\partial^2 F}{\partial x^2}\right)\right)+\frac{1}{E}$$

$$\left(\left(1-\upsilon^2\right)\frac{\partial^2}{\partial x^2}\left(\frac{\partial^2 F}{\partial x^2}\right)-\upsilon\left(1+\upsilon\right)\frac{\partial^2}{\partial x^2}\left(\frac{\partial^2 F}{\partial y^2}\right)\right)-\frac{2\left(1+\upsilon\right)}{E}\frac{\partial^2}{\partial x\,\partial y}\left(-\frac{\partial^2 F}{\partial y\,\partial x}\right)=0$$

by collecting the terms and can be simplified, into the following form:

$$\left(\frac{1-\upsilon^2}{E}\right)\left(\frac{\partial^4 F}{\partial y^4}+\frac{\partial^4 F}{\partial x^4}\right)+\left(\frac{2\left(1+\upsilon\right)-2\upsilon\left(1+\upsilon\right)}{E}\right)\frac{\partial^4 F}{\partial y^2\,\partial x^2}=0$$

It is obvious that $2\left(1+\upsilon\right)-2\upsilon\left(1+\upsilon\right)=2\left(1-\upsilon^2\right)$, and the above equation can be simplified to:

$$\left(\frac{1-\upsilon^2}{E}\right)\left(\frac{\partial^4 F}{\partial y^4}+\frac{\partial^4 F}{\partial x^4}\right)+2\left(\frac{1-\upsilon^2}{E}\right)\frac{\partial^4 F}{\partial y^2\,\partial x^2}=0$$

Then by dividing both sides into $\left(\frac{1-\upsilon^2}{E}\right)$ we have:

$$\frac{\partial^4 F}{\partial y^4}+\frac{\partial^4 F}{\partial x^4}+2\frac{\partial^4 F}{\partial y^2\,\partial x^2}=0 \tag{6.1.28}$$

Which is identical to bi-harmonic equation in Eq. (6.1.22), from this section we can conclude that for both plane stress and plane strain states, the stress function should satisfy the bi-harmonic equation. In the next section we study, the forms of F that can satisfy the Eq. (6.1.28). This enables to express closed form formulas for the stresses.

6.2 DETERMINATION OF F IN TERMS OF COMPLEX VARIABLE FUNCTIONS

In the first section we proved that F satisfies in bi-harmonic equation. Herein we recall the definitions and simplify the equations:

$$\sigma_x = \frac{\partial^2 F}{\partial y^2} \qquad \sigma_y = \frac{\partial^2 F}{\partial x^2}$$

By adding up the two stresses and naming it P, we have:

$$\sigma_x + \sigma_y = \frac{\partial^2 F}{\partial y^2} + \frac{\partial^2 F}{\partial x^2} = \nabla^2 (F) = P \tag{6.2.1}$$

By substituting Eq. (6.2.1) into Eq. (6.1.22), we can change the bi-harmonic equation to Laplace equation as follows:

$$\nabla^2 \left(\nabla^2 (F)\right) = \nabla^2 (P) = 0 \tag{6.2.2}$$

This new function P defined by Eq. (6.2.1) is related function of complex variable as we can see in the following exercise:

Exercise 6.2.1: Prove that P is a real part of a complex function $f(z)$ that is,

$$f(z) = P + iQ \tag{6.2.3}$$

Proof: For any complex function $f(z)$, the real part P and imaginary part Q, should satisfy the Cauchy-Riemann conditions which are:

$$\frac{\partial P}{\partial x} = \frac{\partial Q}{\partial y} \qquad \frac{\partial P}{\partial y} = -\frac{\partial Q}{\partial x} \tag{6.2.4}$$

We differentiate from Eq. (6.2.4) such that we have:

$$\frac{\partial^2 P}{\partial x^2} = \frac{\partial^2 Q}{\partial y \, \partial x} \qquad \frac{\partial^2 P}{\partial y^2} = -\frac{\partial^2 Q}{\partial x \, \partial y} \tag{6.2.5}$$

From Eq. (6.2.5) we can easily find that:

$$\frac{\partial^2 P}{\partial y^2}+\frac{\partial^2 P}{\partial x^2}=\nabla^2(P)=0 \tag{6.2.6}$$

Since Eq. (6.2.2) and Eq. (6.2.6) are identical, our proof is complete. Another type of differentiation from Eq. (6.2.4) gives:

$$\frac{\partial^2 Q}{\partial y^2}=\frac{\partial^2 P}{\partial y\,\partial x} \qquad \frac{\partial^2 Q}{\partial x^2}=-\frac{\partial^2 P}{\partial x\,\partial y} \tag{6.2.7a}$$

From Eq. (6.2.7) we can easily find that:

$$\frac{\partial^2 Q}{\partial y^2}+\frac{\partial^2 Q}{\partial x^2}=\nabla^2(Q)=0 \tag{6.2.7b}$$

Therefore, we can conclude that both real part P and imaginary part Q, of a complex function $f(z)$ will satisfy the Laplace equation and can lead us to the stress function F. These functions (P and Q) are called "harmonic functions" while F is a bi-harmonic function. The next step is defining other complex function $\phi(z)$ defined by:

$$\phi(z)=p+iq=\frac{1}{4}\int f(z)\,dz \tag{6.2.7c}$$

With two features, the first one is:

$$\phi'(z)=\frac{1}{4}f(z) \tag{6.2.8}$$

To find the 2nd feature we differentiate Eq. (6.2.7c) versus x rather than z, and we have:

$$\frac{\partial \phi}{\partial x}=\frac{\partial p}{\partial x}+i\frac{\partial q}{\partial x} \quad \text{and} \quad \frac{\partial \phi}{\partial x}=\frac{\partial \phi}{\partial z}.\frac{\partial z}{\partial x}=\phi'(z) \ \text{ since } \frac{\partial z}{\partial x}=1$$

Also $\phi'(z)=\frac{1}{4}f(z)$ and therefore, we can write:

$$\frac{\partial p}{\partial x}+i\frac{\partial q}{\partial x}=\frac{1}{4}f(z)=\frac{1}{4}(P+iQ)$$

By equating the real parts of the above expression we have:

$$\frac{\partial p}{\partial x}=\frac{1}{4}P \tag{6.2.9}$$

Since p and q are the real and imaginary parts of the complex function $\phi(z)$, from the Cauchy-Riemann conditions, we have:

$$\frac{\partial p}{\partial x}=\frac{\partial q}{\partial y} \tag{6.2.10}$$

Comparing Eq. (6.2.9) and Eq. (6.2.10) gives:

$$\frac{\partial q}{\partial y}=\frac{1}{4}P \tag{6.2.11}$$

So far we know that P, Q, p and q are all harmonic function, while stress function F is bi-harmonic. Koslov in 1909 in his doctoral dissertation found out the relationship between F and the P, Q, p and q as follows here:

$$\begin{aligned}\nabla^2(xp)=\frac{\partial^2}{\partial y^2}(xp)+\frac{\partial^2}{\partial x^2}(xp)&=\frac{\partial}{\partial y}\left(x\frac{\partial p}{\partial y}\right)+\frac{\partial}{\partial x}\left(p+x\frac{\partial p}{\partial x}\right)\\&=x\frac{\partial^2 p}{\partial y^2}+\left(\frac{\partial p}{\partial x}+\frac{\partial p}{\partial x}+x\frac{\partial^2 p}{\partial x^2}\right)\end{aligned}$$

Since p is harmonic then, $\left(\frac{\partial^2 p}{\partial y^2}+\frac{\partial^2 p}{\partial x^2}\right)x=0$, therefore above expression simplifies to:

$$\nabla^2(xp)=2\frac{\partial p}{\partial x} \tag{6.2.12}$$

Exercise 6.2.2: Find an expression for $\nabla^2(yq)$

Solution: similar to the case for $\nabla^2(xp)$ we can write:

$$\nabla^2(yq)=\frac{\partial^2}{\partial y^2}(yq)+\frac{\partial^2}{\partial x^2}(yq)=\frac{\partial}{\partial x}\left(y\frac{\partial q}{\partial x}\right)+\frac{\partial}{\partial y}\left(q+y\frac{\partial q}{\partial y}\right)$$
$$=y\frac{\partial^2 q}{\partial x^2}+\left(\frac{\partial q}{\partial y}+\frac{\partial q}{\partial y}+y\frac{\partial^2 q}{\partial y^2}\right)$$

Since q is harmonic then, $\left(\frac{\partial^2 q}{\partial y^2}+\frac{\partial^2 q}{\partial x^2}\right)y=0$, therefore above expression simplifies to:

$$\nabla^2(yq)=2\frac{\partial q}{\partial y} \tag{6.2.13}$$

We rewrite Eq. (6.2.1) again:

$$\nabla^2(F)=P \tag{6.2.14}$$

Adding up Eqs. (6.2.12), and (6.2.13) together and subtract the result from Eq. (6.2.14) gives:

$$\nabla^2(F)-\nabla^2(xp)-\nabla^2(yq)=P-2\frac{\partial p}{\partial x}-2\frac{\partial q}{\partial y} \tag{6.2.15}$$

We substitute Eqs. (6.2.9) and (6.2.11) into Eq. (6.2.15) which results:

$$\nabla^2(F)-\nabla^2(xp)-\nabla^2(yq)=P-\frac{2}{4}P-\frac{2}{4}P=0 \tag{6.2.16}$$

Laplace operator is a linear one, and mathematically means that:

$$\nabla^2(F)-\nabla^2(xp)-\nabla^2(yq)=\nabla^2(F-xp-yq) \tag{6.2.17}$$

Comparing Eq. (6.2.16) and Eq. (6.2.17) immediately yields to:

$$\nabla^2 (F - x\,p - y\,q) = 0 \tag{6.2.18}$$

From Eq. (6.2.18) we can conclude that, although F is not harmonic, but the combination of $F - x\,p - y\,q$ is a harmonic function. We set this combination to a new function, that is,

$$F - x\,p - y\,q = p_1 \tag{6.2.19}$$

Expressions (6.2.18) and (6.2.19) clearly indicates that p_1 is a harmonic function, and also the stress function F can be expressed in terms of three harmonic functions p, q and p_1 plus two variables x and y like this (see Eq. (6.2.19)):

$$F = p_1 + x\,p + y\,q \tag{6.2.20}$$

Expression (6.2.20) still cannot explain how the stress function F, can be written in terms of complex functions. Now we say that because p_1 is harmonic, it should have a conjugate q_1 which their corresponding complex function $\psi(z)$ is:

$$\psi(z) = p_1 + i\,q_1 \tag{6.2.21}$$

Now we are dealing with four harmonic functions p, q, p_1 and q_1 plus two variables x and y and it seems that situation is more complicated. However, if we look at the definitions (6.2.7) and (6.2.21) and consider that $\bar{z} = x - i\,y$ we can write:

$$\bar{z}\,\phi(z) + \psi(z) = \overbrace{(x - i\,y)}^{\bar{z}}\overbrace{(p + i\,q)}^{\phi(z)} + \overbrace{p_1 + i\,q_1}^{\psi(z)} \tag{6.2.22}$$

In Eq. (6.2.22) we are interested to identify the real part and find out exactly what is it? Can it be related to F?

Exercise 6.2.3: Find an expression for $\Re\left(\bar{z}\,\phi(z)+\psi(z)\right)$.

Solution: From Eq. (6.2.22) we have: $\Re\left(\bar{z}\,\phi(z)+\psi(z)\right)=\Re\left((x-i\,y)(p+i\,q)+p_1+i\,q_1\right)$ leading to: $\Re\left((x-i\,y)(p+i\,q)+p_1+i\,q_1\right)=\Re\left((x\,p+y\,q+p_1)+i(x\,q-y\,p+q_1)\right)=x\,p+y\,q+p_1$ Therefore, it is obvious that:

$$\Re\left(\bar{z}\,\phi(z)+\psi(z)\right)=x\,p+y\,q+p_1 \tag{6.2.23}$$

Comparing Eqs. (6.2.20) and (6.2.23) immediately yields to:

$$F=\Re\left(\bar{z}\,\phi(z)+\psi(z)\right) \tag{6.2.24}$$

From Eq. (6.2.24) we can conclude that the bi-harmonic stress function F, can be expressed by two different analytic complex functions $\phi(z)$ and $\psi(z)$. Moreover, the parameter $\bar{z}$ (which is not analytic) also exist in the expression. This proves that F is not differentiable versus z. The question is can it be differentiated versus x and y? If the answer is yes, then we can calculate the stresses. We discuss about this issue in the next section.

6.3 FORMULAS FOR STRESSES IN TERMS OF COMPLEX FUNCTIONS

In previous section we derived the formula Eq. (6.2.24) for the stress function F. Unfortunately, that form is not directly differentiable, unless we split it into two parts according to this formula:

$$\Re\left(f(z)\right)=\frac{1}{2}\left(f(z)+\overline{f(z)}\right) \tag{6.3.1}$$

Which states that the real part of an analytic function is the sum it and its conjugate dived by two, if we apply this rule to the F in 6.2.24, we have:

$$F=\Re\left(\bar{z}\,\phi(z)+\psi(z)\right)=\frac{1}{2}\left(\bar{z}\,\phi(z)+z\overline{\phi(z)}+\psi(z)+\overline{\psi(z)}\right) \tag{6.3.2}$$

Since we know that $\frac{\partial z}{\partial x}=1$ and also $\frac{\partial \bar{z}}{\partial x}=1$, we can differentiate Eq. (6.3.2) versus x as:

$$\frac{\partial F}{\partial x}=\frac{1}{2}\left(\phi(z)+\bar{z}\phi'(z)+\overline{\phi(z)}+z\overline{\phi'(z)}+\psi'(z)+\overline{\psi'(z)}\right) \quad (6.3.3)$$

Exercise 6.3.1: Find out what is $\frac{\partial F}{\partial y}$?

Solution: we have $\frac{\partial z}{\partial y}=i$ and also $\frac{\partial \bar{z}}{\partial y}=-i$, and can differentiate Eq. (6.3.2) versus y as:

$$\frac{\partial F}{\partial y}=\frac{1}{2}i\left(-\phi(z)+\bar{z}\phi'(z)+\overline{\phi(z)}-z\overline{\phi'(z)}+\psi'(z)-\overline{\psi'(z)}\right) \quad (6.3.4)$$

It should be remembered that for obtaining (6.3.4) we have used this rule:

$$\overline{f'(z)}=\frac{\partial \overline{f(z)}}{\partial \bar{z}} \Rightarrow \frac{\partial \overline{f(z)}}{\partial y}=\overline{f'(z)}\frac{\partial \bar{z}}{\partial y}=-i\overline{f'(z)}$$

If we multiply Eq. (6.3.4) by i and add up the result to Eq. (6.3.3) then we have:

$$\frac{\partial F}{\partial x}+i\frac{\partial F}{\partial y}=\left(\phi(z)+z\overline{\phi'(z)}+\overline{\psi'(z)}\right) \quad (6.3.5)$$

We will use Eq. (6.3.5) later, but for the stresses we need to differentiate Eq. (6.3.5) again versus x knowing that $\frac{\partial z}{\partial x}=1$ and also $\frac{\partial \bar{z}}{\partial x}=1$. This yields to:

$$\frac{\partial^2 F}{\partial x^2}+i\frac{\partial^2 F}{\partial x\,\partial y}=\left(\phi'(z)+\overline{\phi'(z)}+z\overline{\phi''(z)}+\overline{\psi''(z)}\right) \quad (6.3.6)$$

Exercise 6.3.2: What is $i\dfrac{\partial^2 F}{\partial y^2}+\dfrac{\partial^2 F}{\partial y\,\partial x}$?

Solution: we have $\dfrac{\partial z}{\partial y}=i$ and also $\dfrac{\partial \bar{z}}{\partial y}=-i$, and can differentiate Eq. (6.3.5) versus y as:

$$i\frac{\partial^2 F}{\partial y^2}+\frac{\partial^2 F}{\partial x\,\partial y}=i\left(\phi'(z)+\overline{\phi'(z)}-z\overline{\phi''(z)}-\overline{\psi''(z)}\right) \tag{6.3.7}$$

If we multiply both sides of Eq. (6.3.7) into i then we have:

$$-\frac{\partial^2 F}{\partial y^2}+i\frac{\partial^2 F}{\partial x\,\partial y}=\left(z\overline{\phi''(z)}+\overline{\psi''(z)}-\phi'(z)-\overline{\phi'(z)}\right) \tag{6.3.8}$$

If we take away Eq. (6.3.8) from Eq. (6.3.6) the result will be:

$$\frac{\partial^2 F}{\partial y^2}+\frac{\partial^2 F}{\partial x^2}=2\left(\phi'(z)+\overline{\phi'(z)}\right)$$

If we substitute Eq. (6.2.1) into above equation and use the Eq. (6.3.1) then we have:

$$\sigma_x+\sigma_y=4\Re\left(\phi'(z)\right) \tag{6.3.9}$$

We can conclude the Eq. (6.3.9) easily by comparing Eqs. (6.2.1), (6.2.3) and (6.2.8) from the previous section, so derivation of Eq. (6.3.9) can be confirmed. However, extra information can be picked up by adding up the Eqs. (6.3.6) and (6.3.8) such that:

$$\frac{\partial^2 F}{\partial x^2}-\frac{\partial^2 F}{\partial y^2}+2i\frac{\partial^2 F}{\partial x\,\partial y}=2\left(z\overline{\phi''(z)}+\overline{\psi''(z)}\right) \tag{6.3.10}$$

If we substitute the equations in Eq. (6.1.21) into the Eq. (6.3.10) we obtain a formula in terms of stresses like this:

$$\sigma_y-\sigma_x+2i\tau_{xy}=2\left(z\overline{\phi''(z)}+\overline{\psi''(z)}\right) \tag{6.3.11}$$

For finding the conjugate form of Eq. (6.3.11) we change i with $-i$, in the left side and in the right side we change like this:

$$\sigma_y - \sigma_x - 2i\tau_{xy} = 2\left(\bar{z}\phi''(z) + \psi''(z)\right) \tag{6.3.12}$$

By adding up the Eqs. (6.3.11) and (6.3.12) we have:

$$2\left(\sigma_y - \sigma_x\right) = 2\left(\bar{z}\phi''(z) + z\overline{\phi''(z)} + \psi''(z) + \overline{\psi''(z)}\right)$$

Which by using the Eq. (6.3.1), can be simplified to:

$$\sigma_y - \sigma_x = 2\Re\left(\bar{z}\phi''(z) + \psi''(z)\right) \tag{6.3.13}$$

Adding up Eq. (6.3.9) and Eq. (6.3.13) and halve the result provides σ_y as:

$$\sigma_y = 2\Re\left(\phi'(z)\right) + \Re\left(\bar{z}\phi''(z) + \psi''(z)\right) \tag{6.3.14}$$

Taking away Eq. (6.3.13) from Eq. (6.3.9) and halve the result provides σ_x as:

$$\sigma_x = 2\Re\left(\phi'(z)\right) - \Re\left(\bar{z}\phi''(z) + \psi''(z)\right) \tag{6.3.15}$$

Exercise 6.3.3: what is the expression for τ_{xy}?

Solution: By taking away Eq. (6.3.12) from Eq. (6.3.11) and dividing the result by 4 we have:

$$i\tau_{xy} = \frac{1}{2}\left(\bar{z}\phi''(z) - \underbrace{z\overline{\phi''(z)}}_{conjugate} + \psi''(z) - \underbrace{\overline{\psi''(z)}}_{conjugate}\right)$$

From complex variable theory we have, $f(z) - \overline{f(z)} = 2i\Im\left(f(z)\right)$. This is very easy to verify, then we can use this formula for the above expression and we have, $i\tau_{xy} = i\Im\left(\bar{z}\phi''(z) + \psi''(z)\right)$ which yields to:

$$\tau_{xy} = \Im\left(\bar{z}\phi''(z) + \psi''(z)\right) \tag{6.3.16}$$

The stresses that produced near the boundary of holes or cracks, depends on the external loads and moments that applied to face of the hole and crack. We consider an element near the boundary shown in Fig. 6.1. If the thickness of the element is $h=1$ and the X is the boundary stress in x direction and Y is the boundary stress in y direction, the equilibrium according to Fig. 6.1 is:

$$\sum F_x = 0 \Rightarrow X \underbrace{ds.1}_{area} = \sigma_x \underbrace{dy.1}_{area} + \tau_{xy} \underbrace{-dx.1}_{area}$$

According to Fig. 6.1, near the boundary when $dy > 0$, by moving across ds, then $dx < 0$ and therefore the area $\underbrace{-dx.1}_{area}$ will be positive. If we divide both sides of the above equation into ds then we have:

$$X = \frac{dy}{ds}\sigma_x - \frac{dx}{ds}\tau_{xy} \tag{6.3.17}$$

Similarly in y direction the equilibrium leads to:

$$\sum F_y = 0 \Rightarrow Y \underbrace{ds.1}_{area} = \sigma_y \underbrace{-dx.1}_{area} + \tau_{xy} \underbrace{dy.1}_{area}$$

Again we divide both sides of the above equation into ds then we have:

$$Y = -\frac{dx}{ds}\sigma_x + \frac{dy}{ds}\tau_{xy} \tag{6.3.18}$$

If from Eq. (6.1.21) we substitute the stress functions into Eqs. (6.3.17) and (6.3.19) we have:

$$X = \frac{\partial^2 F}{\partial y^2}\frac{dy}{ds} + \frac{\partial^2 F}{\partial x \partial y}\frac{dx}{ds} \qquad Y = -\frac{\partial^2 F}{\partial x^2}\frac{dx}{ds} - \frac{\partial^2 F}{\partial x \partial y}\frac{dy}{ds}$$

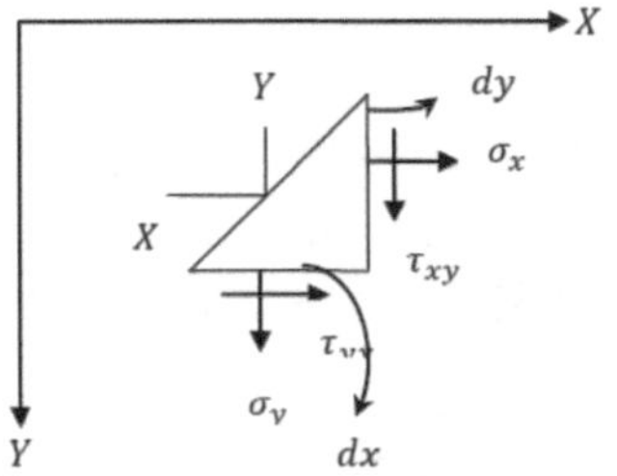

FIGURE 6.1 Element near the boundary.

Rewriting the above equation into this form:

$$X = \frac{\partial}{\partial y}\left(\frac{\partial F}{\partial y}\right)\frac{dy}{ds} + \frac{\partial}{\partial x}\left(\frac{\partial F}{\partial y}\right)\frac{dx}{ds} \qquad Y = -\frac{\partial}{\partial x}\left(\frac{\partial F}{\partial x}\right)\frac{dx}{ds} - \frac{\partial}{\partial y}\left(\frac{\partial F}{\partial x}\right)\frac{dy}{ds} \tag{6.3.19}$$

The forms shown in Eq. (6.3.19) are complete differential forms that are defined by:

$$\frac{d}{ds} = \frac{\partial}{\partial x}\frac{dy}{ds} + \frac{\partial}{\partial y}\frac{dx}{ds} \tag{6.3.20}$$

The above operator (6.3.20) is also called the chain rule in differentiation. If we simplify the expressions in Eq. (6.3.19) according to the chain rule in Eq. (6.3.20) we have:

$$X = \frac{d}{ds}\left(\frac{\partial F}{\partial y}\right) \tag{6.3.21}$$

$$Y = -\frac{d}{ds}\left(\frac{\partial F}{\partial x}\right) \tag{6.3.22}$$

If we multiply both sides of Eqs. (6.3.21) and (6.3.22) into ds and the integrate we have:

$$f_x = \int_A^B X\,ds = \left(\frac{\partial F}{\partial y}\right)_A^B \tag{6.3.23}$$

$$f_y = \int_A^B Y\,ds = -\left(\frac{\partial F}{\partial x}\right)_A^B \tag{6.3.24}$$

It should be remembered that the integration starts from point A in the boundary to the point B if the boundary is closed curve then starts from

A and returns to A. If we multiply Eq. (6.3.24) into i and add up with Eq. (6.3.23) then we have:

$$f_x + i f_y = \left(\frac{\partial F}{\partial y} - i \frac{\partial F}{\partial x} \right)$$

This can be written into an alternative form like:

$$f_x + i f_y = -i \left(\frac{\partial F}{\partial x} + i \frac{\partial F}{\partial y} \right) \tag{6.3.25}$$

IIf we compare the Eqs. (6.3.5) and (6.3.25), then we can easily conclude that:

$$f_x + i f_y = -i \left(\phi(z) + z\overline{\phi'(z)} + \overline{\psi'(z)} \right)_A^B \tag{6.3.26}$$

The Eq. (6.3.26) will be used in incoming articles and is important in both elasticity and fracture mechanic theories.

6.4 EXPRESSION FOR DISPLACEMENT IN TERMS OF COMPLEX FUNCTIONS

In fracture mechanics we are interested to know how much the crack faces are opening when the load is applied. In other branches of solid mechanics the displacement may not be that important. Fist we need to substitute the strain formulas in Eq. (6.1.1) into the Eq. (6.1.8), (6.1.9), and (6.1.10) they could expressed into this form:

$$E \frac{\partial u}{\partial x} = \sigma_x - \upsilon \sigma_y \tag{6.4.1}$$

$$E \frac{\partial v}{\partial y} = \sigma_y - \upsilon \sigma_x \tag{6.4.2}$$

$$\frac{E}{2(1+\upsilon)} \left(\frac{\partial u}{\partial y} + \frac{\partial v}{\partial x} \right) = \tau_{xy} \tag{6.4.3}$$

We substitute the stress functions defined by Eq. (6.1.21) into Eqs. (6.4.1), and (6.4.2) we have:

$$E\frac{\partial u}{\partial x}=\frac{\partial^2 F}{\partial y^2}-\upsilon\frac{\partial^2 F}{\partial x^2} \tag{6.4.4}$$

$$E\frac{\partial v}{\partial y}=\frac{\partial^2 F}{\partial x^2}-\upsilon\frac{\partial^2 F}{\partial y^2} \tag{6.4.5}$$

It is necessary to express Eq. (6.4.4) in terms of x and Eq. (6.4.5) in terms of y and to do this we use Eq. (6.2.1) to involve P in the above equations and therefore we have:

$$E\frac{\partial u}{\partial x}=\left(P-\frac{\partial^2 F}{\partial x^2}\right)-\upsilon\frac{\partial^2 F}{\partial x^2}=-(1+\upsilon)\frac{\partial^2 F}{\partial x^2}+P \tag{6.4.6}$$

$$E\frac{\partial v}{\partial y}=\left(P-\frac{\partial^2 F}{\partial y^2}\right)-\upsilon\frac{\partial^2 F}{\partial y^2}=-(1+\upsilon)\frac{\partial^2 F}{\partial y^2}+P \tag{6.4.7}$$

To change Eq. (6.4.6) we use Eq. (6.2.9) and for changing Eq. (6.4.7) we use Eq. (6.2.11) to replace P such that:

$$E\frac{\partial u}{\partial x}==-(1+\upsilon)\frac{\partial^2 F}{\partial x^2}+4\frac{\partial p}{\partial x} \tag{6.4.8}$$

$$E\frac{\partial v}{\partial y}=-(1+\upsilon)\frac{\partial^2 F}{\partial y^2}+4\frac{\partial q}{\partial y} \tag{6.4.9}$$

By integrating from Eq. (6.4.8) versus x and integrating Eq. (6.4.9) versus y we have:

$$Eu=-(1+\upsilon)\frac{\partial F}{\partial x}+4p+A(y) \tag{6.4.10}$$

$$E\,v = -(1+\upsilon)\frac{\partial F}{\partial y} + 4q + B(x) \tag{6.4.11}$$

We need to find out $A(y)$ and also $B(x)$ to have formula for u and v. To do this, we differentiate Eq. (6.4.10) versus y and Eq. (6.4.11) versus x for having this:

$$E\frac{\partial u}{\partial y} = -(1+\upsilon)\frac{\partial^2 F}{\partial y\,\partial x} + 4\frac{\partial p}{\partial y} + \frac{d\,A(y)}{dy} \text{ which results:}$$

$$\frac{\partial u}{\partial y} = -\frac{(1+\upsilon)}{E}\frac{\partial^2 F}{\partial y\,\partial x} + \frac{4}{E}\frac{\partial p}{\partial y} + \frac{1}{E}\frac{d\,A(y)}{dy} \tag{6.4.12}$$

$$E\frac{\partial v}{\partial x} = -(1+\upsilon)\frac{\partial^2 F}{\partial y\,\partial x} + 4\frac{\partial q}{\partial x} + \frac{d\,B(x)}{dx} \text{ which results:}$$

$$\frac{\partial v}{\partial x} = -\frac{(1+\upsilon)}{E}\frac{\partial^2 F}{\partial y\,\partial x} + \frac{4}{E}\frac{\partial q}{\partial x} + \frac{1}{E}\frac{d\,B(x)}{dx} \tag{6.4.13}$$

Then we substitute Eqs. (6.4.12) and (6.4.13) into Eq. (6.4.3) which provides a long expression:

$$\frac{E}{2(1+\upsilon)}\begin{pmatrix} -\dfrac{(1+\upsilon)}{E}\dfrac{\partial^2 F}{\partial y\,\partial x} + \dfrac{4}{E}\dfrac{\partial p}{\partial y} + \dfrac{1}{E}\dfrac{d\,A(y)}{dy} \\ -\dfrac{(1+\upsilon)}{E}\dfrac{\partial^2 F}{\partial y\,\partial x} + \dfrac{4}{E}\dfrac{\partial q}{\partial x} + \dfrac{1}{E}\dfrac{d\,B(x)}{dx} \end{pmatrix} = \tau_{xy}$$

and can be simplified to:

$$-\frac{\partial^2 F}{\partial y\,\partial x} + \frac{2}{1+\upsilon}\left(\frac{\partial p}{\partial y} + \frac{\partial q}{\partial x}\right) + \frac{1}{2(1+\upsilon)}\left(\frac{d\,A(y)}{dy} + \frac{d\,B(x)}{dx}\right) = \tau_{xy}$$

In the above formula $\tau_{xy} = -\frac{\partial^2 F}{\partial y \partial x}$ and since p, q are components of $\phi(z)$ in 6.2.7 and satisfy Cauchy-Riemann condition, that is, $\frac{\partial p}{\partial y} = -\frac{\partial q}{\partial x}$ then it reduces to:

$$\frac{dA(y)}{dy} + \frac{dB(x)}{dx} = 0 \Rightarrow \frac{dA(y)}{dy} = -\frac{dB(x)}{dx} + C$$

The integration leads to $A(y) = Cy + C_1$ which contribute to u and $B(x) = -Cx + C_2$ which contribute to v. They cannot contribute to stress functions because their 2nd derivatives are zero. They are not produced by applying the load and we can express them as rigid body motions, therefore it can be ignored and then Eqs. (6.4.10) and (6.4.11) will be simplified to:

$$u = -\frac{(1+\upsilon)}{E}\frac{\partial F}{\partial x} + \frac{4}{E}p \tag{6.4.14}$$

$$v = -\frac{(1+\upsilon)}{E}\frac{\partial F}{\partial y} + \frac{4}{E}q \tag{6.4.15}$$

If we multiply Eq. (6.4.15) to i and add the result to the Eq. (6.4.14) then we have:

$$u + iv = -\frac{(1+\upsilon)}{E}\left(\frac{\partial F}{\partial x} + i\frac{\partial F}{\partial y}\right) + \frac{4}{E}(p + iq)$$

In the above equation instead of $p + iq$ we substitute from 6.2.7 and for $\frac{\partial F}{\partial x} + i\frac{\partial F}{\partial y}$ from 6.3.5 then u and can be expressed by complex functions as:

$$u + iv = -\frac{(1+\upsilon)}{E}\left(\phi(z) + z\overline{\phi'(z)} + \overline{\psi'(z)}\right) + \frac{4}{E}\phi(z)$$

Which can be simplified to:

$$u+iv=\frac{3-\upsilon}{E}\phi(z)-\frac{1+\upsilon}{E}\left(\phi(z)+z\overline{\phi'(z)}+\overline{\psi'(z)}\right) \qquad (6.4.16)$$

The Eq. (6.4.16) is very important and we will use it later. It is valid for the plane stress case only. For the plane strain case something like Eq. (6.4.17) is needed. It is obvious that for plane strain case the Eqs. (6.4.1)–(6.4.9) should be changed. For this purpose we return to Eqs. (6.1.25) and (6.1.26) and write them in terms of the displacement into this form:

$$E\frac{\partial u}{\partial x}=\left(1-\upsilon^2\right)\sigma_x-\upsilon\left(1+\upsilon\right)\sigma_y \qquad (6.4.17)$$

$$E\frac{\partial v}{\partial y}=\left(1-\upsilon^2\right)\sigma_y-\upsilon\left(1+\upsilon\right)\sigma_x \qquad (6.4.18)$$

Exercise 6.4.1: Demonstrate integration of Eqs. (6.4.17) and (6.4.18).

Solution: By substitution of the stress function from Eq. (6.1.21) into Eqs. (6.4.17) and (6.4.18) we have:

$$E\frac{\partial u}{\partial x}=\left(1-\upsilon^2\right)\frac{\partial^2 F}{\partial y^2}-\upsilon\left(1+\upsilon\right)\frac{\partial^2 F}{\partial x^2} \qquad E\frac{\partial v}{\partial y}=\left(1-\upsilon^2\right)\frac{\partial^2 F}{\partial x^2}-\upsilon\left(1+\upsilon\right)\frac{\partial^2 F}{\partial y^2}$$

Again by using Eq. (6.2.1) we convert the equations into P as follows:

$$E\frac{\partial u}{\partial x}=\left(1-\upsilon^2\right)\left(P-\frac{\partial^2 F}{\partial x^2}\right)-\upsilon\left(1+\upsilon\right)\frac{\partial^2 F}{\partial x^2}$$

$$E\frac{\partial v}{\partial y}=\left(1-\upsilon^2\right)\left(P-\frac{\partial^2 F}{\partial x^2}\right)-\upsilon\left(1+\upsilon\right)\frac{\partial^2 F}{\partial y^2}$$

They can be simplified into:

$$E\frac{\partial u}{\partial x}=-\left(1+\upsilon\right)\frac{\partial^2 F}{\partial x^2}+\left(1-\upsilon^2\right)P \qquad (6.4.19)$$

$$E\frac{\partial v}{\partial y}=-(1+\upsilon)\frac{\partial^2 F}{\partial y^2}+(1-\upsilon^2)\,P \tag{6.4.20}$$

For the plain strain case Eqs. (6.4.19) and (6.4.20) are applicable, and the Eqs. (6.4.8) and (6.4.9) changes to (use Eq. (6.2.9) and also Eq. (6.2.11)):

$$E\frac{\partial u}{\partial x}=-(1+\upsilon)\frac{\partial^2 F}{\partial x^2}+4(1-\upsilon^2)\frac{\partial p}{\partial x} \tag{6.4.21}$$

$$E\frac{\partial v}{\partial y}=-(1+\upsilon)\frac{\partial^2 F}{\partial y^2}+4(1-\upsilon^2)\frac{\partial q}{\partial y} \tag{6.4.22}$$

Through integration of Eqs. (6.4.21) and (6.4.22) it is needless to say that $A(y)$ and $B(x)$ appear again but they will be ignored as we discussed. Therefore, the new forms of the Eqs. (6.4.14) and (6.4.15) for plain strain situation will be:

$$u=-\frac{(1+\upsilon)}{E}\frac{\partial F}{\partial x}+\frac{4(1-\upsilon^2)}{E}p \tag{6.4.23}$$

$$v=-\frac{(1+\upsilon)}{E}\frac{\partial F}{\partial y}+\frac{4(1-\upsilon^2)}{E}q \tag{6.4.24}$$

If we multiply Eq. (6.4.24) to i and add the result to the Eq. (6.4.23) then we have:

$$u+iv=-\frac{(1+\upsilon)}{E}\left(\frac{\partial F}{\partial x}+i\frac{\partial F}{\partial y}\right)+\frac{4(1-\upsilon^2)}{E}(p+iq)$$

In the above equation instead of $p+iq$ we substitute from Eq. (6.2.7) and for $\frac{\partial F}{\partial x}+i\frac{\partial F}{\partial y}$ from Eq. (6.3.5) then u and can be expressed by complex functions as:

$$u+iv=-\frac{(1+\upsilon)}{E}\left(\phi(z)+z\overline{\phi'(z)}+\overline{\psi'(z)}\right)+\frac{4(1-\upsilon^2)}{E}\phi(z)$$

Which can be simplified to:

$$u+iv=\frac{3-\upsilon-4\upsilon^2}{E}\phi(z)-\frac{1+\upsilon}{E}\left(z\overline{\phi'(z)}+\overline{\psi'(z)}\right)$$

Since $3-\upsilon-4\upsilon^2=(3-4\upsilon)(1+\upsilon)$ the above equation is:

$$u+iv=\frac{(3-4\upsilon)(1+\upsilon)}{E}\phi(z)-\frac{1+\upsilon}{E}\left(z\overline{\phi'(z)}+\overline{\psi'(z)}\right) \quad (6.4.25)$$

The Eqs. (6.4.16) and (6.4.25) provides the expressions for the displacements for plane stress and plane strain, respectively. The can be abbreviated to one formula according to these definitions given in many references.

$$2\mu(u+iv)=\kappa\phi(z)-z\overline{\phi'(z)}-\overline{\psi'(z)} \quad (6.4.26)$$

$$\mu=\frac{E}{2(1+\upsilon)} \qquad \kappa=\begin{cases}\dfrac{3-\upsilon}{1+\upsilon} & plane\,stress\\ 3-4\upsilon & plane\,strain\end{cases} \quad (6.4.27)$$

6.5 STRUCTURE OF THE FUNCTIONS $\phi(z)$ AND $\psi(z)$

In this section we study the functions as result of following remote loadings:

In the above Fig. 6.2 in case (a) the remote stress is $\sigma_y=\sigma$ and in case (b) the remote stress is $\sigma_x=\sigma$ and in (c) the shear stress is $\tau_{xy}=\tau$.

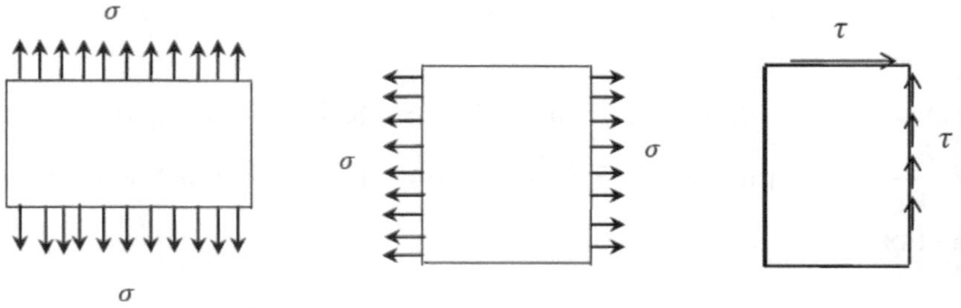

FIGURE 6.2 (a) the remote stress is $\sigma_y=\sigma$; (b) the remote stress is $\sigma_x=\sigma$; and (c) the shear stress is $\tau_{xy}=\tau$

The loading may also be the combinations of the above. The question is what kind of functions $\phi(z)$ and $\psi(z)$ can match the above remote stresses? The answer to this question lies on the Eq. (6.3.9) that says:

$$\sigma_x + \sigma_y = 4\Re\left(\phi'(z)\right)$$

The remote stress that is, in $z \to \infty$ has a bounded value since they are the loads are σ and τ, then it is necessary that $\phi'(z) = C$ is constant value. Therefore, the functions $\phi(z)$ and $\psi(z)$ should take the following forms:

$$\phi(z) = Az \tag{6.5.1}$$

$$\psi'(z) = Bz \tag{6.5.2}$$

By substituting Eqs. (6.5.1) and (6.5.2) into Eqs. (6.3.14) and (6.3.15) we have:

$\sigma_y = 2\Re\left(\phi'(z)\right) + \Re\left(\bar{z}\phi''(z) + \psi''(z)\right) \Rightarrow \sigma = 2\Re(A) + \Re(B)$ since in $z \to \infty$ the stress is $\sigma_y = \sigma$ and then we have:

$$2A + B = \sigma \tag{6.5.3}$$

$\sigma_x = 2\Re\left(\phi'(z)\right) - \Re\left(\bar{z}\phi''(z) + \psi''(z)\right) \Rightarrow 0 = 2\Re(A) - \Re(B)$ since in $z \to \infty$ the stress is $\sigma_x = 0$ and then we have:

$$2A - B = 0 \tag{6.5.4}$$

From Eqs. (6.5.3) and (6.5.4) it is obvious that $A = \dfrac{\sigma}{4}$ and $B = \dfrac{\sigma}{2}$, therefore it will change to:

$$\phi(z) = \frac{\sigma}{4} z \tag{6.5.5}$$

$$\psi'(z) = \frac{\sigma}{2} z \tag{6.5.6}$$

Loading of a crack in case (b) even if it happens is not dangerous, but in load case (c) for shear stress we need to substitute, Eqs. (6.5.1) and (6.5.2) into Eq. (6.3.16) then we have:
$\tau_{xy} = \Im\left(\bar{z}\phi''(z)+\psi''(z)\right) \Rightarrow \tau = \Im(B)$ since in $z \to \infty$ the stress is $\tau_{xy} = \tau$ and then we have:

$$\Im(B) = \tau \tag{6.5.7}$$

Since in this case $\sigma_y = 0$, then we have: $2\Re(A) + \Re(B) = 0$, and if multiple Eq. (6.5.7) by and add it up to this we have: $2\Re(A) + \Re(B) + i\Im(B) = i\tau$. Moreover, $\sigma_x = 0$ which yields to $2\Re(A) - \Re(B) = 0$, and we can conclude that both $\Re(B) = 0$ and also $\Re(A) = 0$, this means that $B = i\tau$ without real part. This yields to:

$$\psi'(z) = i\tau z \tag{6.5.8}$$

Now the question is what is the $\Im(A)$? We will answer to this later, which means that $\phi(z)$ cannot be determined yet. It should be remembered that Eqs. (6.5.5), (6.5.6) and (6.5.8) are valid for a body without any hole or crack. In presence of a crack or hole, Eqs. (6.5.5) and (6.5.6) will not be valid unless we study in a region very far from the crack or hole. In this case Eqs. (6.5.5) and (6.5.6) changes to:

$$\underset{z\to\infty}{\varphi(z)} = \frac{\sigma}{4} z \tag{6.5.9}$$

$$\underset{z\to\infty}{\psi'(z)} = \frac{\sigma}{2} z \tag{6.5.10}$$

Obviously we need to find $\psi'(z)$ and also $\phi(z)$, in the region near the crack or hole. For this purpose first it is necessary to describe that region near the flaw, mathematically. We can express a flaw or crack as an ellipse with major semi.avis a and minor semi.avis b in which $a >> b$ or $\frac{b}{a} \to \infty$, the shaded area outside the flaw is shown in the Fig.6.3, for this ellipse we define the eccentricity as:

$$m = \frac{a-b}{a+b} \tag{6.5.11}$$

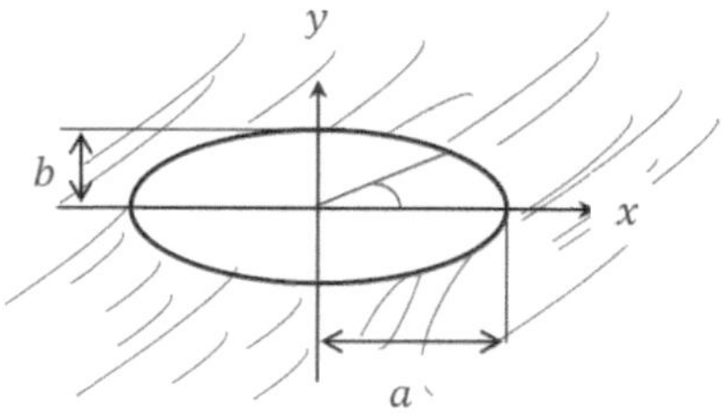

FIGURE 6.3 z plane

And the average diameter by:

$$c = \frac{a+b}{2} \tag{6.5.12}$$

In Fig. 6.3, the area outside the ellipse can be simplified by mapping it into outside of a unit circle, represented by a shaded area in Fig. 6.4. The area outside the ellipse

can be expressed by z while the area outside a unit circle is represented by ζ and mathematically can be expressed by:

$$\zeta = \rho e^{i\theta} \quad \rho \geq 1 \tag{6.5.13}$$

In complex variable theory it can be proved that with an appropriate mapping function, we can map the region outside an ellipse into a corresponding region outside a unit circle we call it w plane. The mapping function is:

$$z = w(\zeta) = \left(\zeta + \frac{m}{\zeta}\right)c \tag{6.5.14}$$

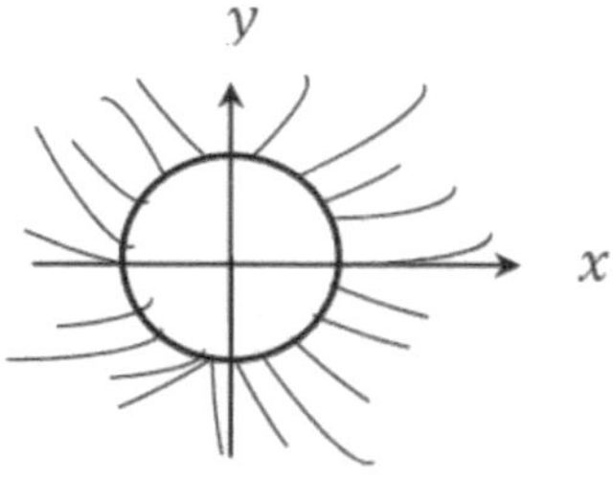

FIGURE 6.4 w plane

Although Eq. (6.5.14) can be proved in complex variable theory, herein we can demonstrate how this mapping can occur. It is necessary to substitute ζ from Eq. (6.5.13) into Eq. (6.5.14) so that:

$z=\left(\rho e^{i\theta}+\frac{m}{\rho}e^{-i\theta}\right)c=\left(\rho(\cos\theta+i\sin\theta)+\frac{m}{\rho}(\cos\theta-i\sin\theta)\right)c$ that can be simplified as:

$$z=\left(\left(\rho+\frac{m}{\rho}\right)\cos\theta+i\left(\rho-\frac{m}{\rho}\right)\sin\theta\right)c$$

If we want to know what unit circle itself describes? In the above expression we substitute $\rho=1$ and then followed by Eqs. (6.5.11) and (6.5.12) we have:

$$z=\left((1+m)\cos\theta+i(1-m)\sin\theta\right)c=\begin{pmatrix}\left(1+\dfrac{a-b}{a+b}\right)\cos\theta+\\ i\left(1-\dfrac{a-b}{a+b}\right)\sin\theta\end{pmatrix}\frac{a+b}{2}$$

identical to:

$z=a\cos\theta+ib\sin\theta$ also in z plane we have $z=x+iy$ and this results $x=a\cos\theta$ and $y=b\sin\theta$ which describes the ellipse $\frac{x^2}{a^2}+\frac{y^2}{b^2}=1$. This explains how an ellipse maps to an unit circle and the area outside it maps into outside of the unit circle. Our objective is that mapping the equation Eq. (6.3.26) from the z plane to the w plane, by using the above transformation. We start by multiply both sides of the Eq. (6.3.26) into *i*, then:

$$if_x-f_y=\left(\phi(z)+z\overline{\phi'(z)}+\overline{\psi'(z)}\right)_A^B \quad (6.5.15)$$

For expressing Eq. (6.5.15) in w plane (versus ζ) we need to remember that since $z=w(\zeta)$, we have $\phi(z)=\phi(w(\zeta))=\phi(\zeta)$, therefore in differentiation we need to follow the chain rule that is, $\phi'(z)=\frac{d\phi}{dz}=\frac{d\phi}{d\zeta}\frac{d\zeta}{dz}=$

$\frac{d\phi}{d\zeta}\left(\frac{dz}{d\zeta}\right)^{-1}$, and since $\frac{dz}{d\zeta}=w'(\zeta)$ we can write, $\phi'(z)=\frac{\phi'(\zeta)}{w'(\zeta)}$ for the other term in Eq. (6.5.15) see the following exercise:

Exercise 6.5.1: Find an expression for $z\overline{\phi'(z)}$ in w plane:

Solution: it is easy to verify that for two complex numbers A and B, we have $\overline{\left(\frac{A}{B}\right)}=\frac{\bar{A}}{\bar{B}}$ and we can apply this rule to $\overline{\phi'(z)}$ can show that $z\overline{\phi'(z)}=w(\zeta)\frac{\overline{\phi'(\zeta)}}{\overline{w'(\zeta)}}$.

For the function $\psi'(z)$ in Eq. (6.5.15) we define,

$$\psi_I(z)=\psi'(z) \tag{6.5.16}$$

So that we can have, $\overline{\psi'(z)}=\overline{\psi_I(z)}\Rightarrow\overline{\psi_I(w(\zeta))}=\overline{\psi_I(\zeta)}$. Now it is possible to express Eq. (6.5.15) in w plane but it should be remembered that (6.5.15) is valid only on the ellipse but not outside of it. If we want to interpret this in w plane, it means that instead of z in Eq. (6.5.15) we cannot substitute $\zeta=\rho e^{i\theta}$ but we need to substitute $z=\eta=e^{i\theta}$ which represents the ellipse $\frac{x^2}{a^2}+\frac{y^2}{b^2}=1$, this is: $i\,f_x-f_y=\left(\phi(\eta)+w(\eta)\frac{\overline{\phi'(\eta)}}{\overline{w'(\eta)}}+\overline{\psi_I(\eta)}\right)_A^B$

and the conjugate of this is:

$$-i\,f_x-f_y=\left(\overline{\phi(\eta)}+\overline{w(\eta)}\frac{\phi'(\eta)}{w'(\eta)}+\psi_I(\eta)\right)_A^B=0 \tag{6.5.17}$$

The Eq. (6.5.17) is applicable in fracture mechanics, where the faces of the crack or flaw, is not loaded. We can integrate Eq. (6.5.17) over the closed circular contour with unit radius named γ that can be expressed by $\eta=e^{i\theta}$, so that the integral is also zero that is,

$$\int_\gamma\psi_I(\eta)d\eta+\int_\gamma\overline{w(\eta)}\frac{\phi'(\eta)}{w'(\eta)}d\eta+\int_\gamma\overline{\phi(\eta)}d\eta=0 \tag{6.5.18}$$

In Eq. (6.5.18) $\eta = e^{i\theta}$ and obviously $\bar{\eta} = e^{-i\theta} = \frac{1}{\eta}$. Since the stresses are finite, the functions $\psi_I(\zeta)$ and $\phi(\zeta)$, and followed by that $\psi_I(\eta)$ and $\phi(\eta)$, should have the following forms:

$$\psi_I(\eta) = A_1\eta + B_1/\eta + B_2/\eta^2 + \cdots + B_n/\eta^n$$

$$\overline{\phi(\eta)} = C_1\eta + D_1/\eta + D_2/\eta^2 + \cdots + D_n/\eta^n$$

$$\overline{w(\eta)}\frac{\phi'(\eta)}{w'(\eta)} = E_1\eta + F_1/\eta + F_2/\eta^2 + \cdots + F_n/\eta^n$$

All of the above functions at $\eta = 0$ are singular and therefore they are not analytic inside the unit circle γ, which means that the integrals in Eq. (6.5.18) cannot be evaluated. The alternative suggestion is that multiply both sides of Eq. (6.5.17) into $\frac{1}{\eta - \zeta}$ and then integrate also remembering that $\overline{w(\eta)} = w(\bar{\eta}) = w(\eta^{-1})$ so that:

$$\int_\gamma \frac{\psi_I(\eta)}{\eta - \zeta} d\eta + \int_\gamma \frac{w(\eta^{-1})}{\eta - \zeta}\frac{\phi'(\eta)}{w'(\eta)} d\eta + \int_\gamma \frac{\phi(\eta^{-1})}{\eta - \zeta} d\eta = 0 \qquad (6.5.19)$$

The integrals in Eq. (6.5.19) at $\eta = \zeta$ which is outside the unit circle are also singular and the stresses cannot be determined. The only way is changing the integration contour from γ to another contour Γ which is shown in Fig. 6.5. According to the Fig. 6.5, the points $\eta = \zeta$ and $\eta = 0$ are outside the contour Γ, and it consists of several paths that is,

$$\Gamma = \gamma + L_1 + L_2 + L_3 + L_4 + L_5 + L_6 + \Gamma_1$$

The integrals in Eq. (6.5.19) are all analytic in Γ and therefore, Cauchy integral formula can be implemented, that is,

$$\int_\Gamma \frac{\psi_I(\eta)}{\eta - \zeta} d\eta = 0 \qquad (6.5.20)$$

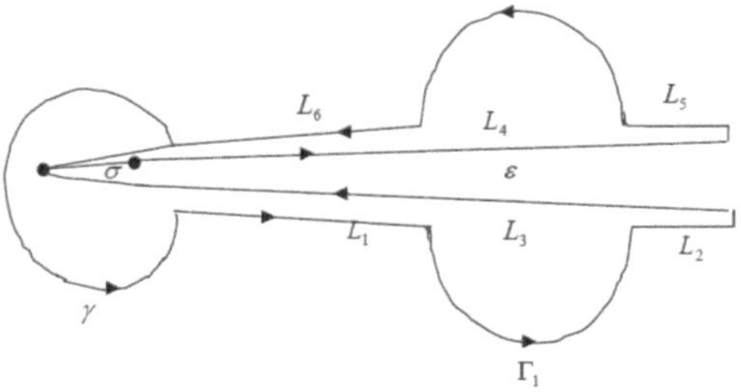

FIGURE 6.5 closed contour Γ.

$$\int_{\Gamma} \frac{w(\eta^{-1})}{\eta-\zeta} \frac{\phi'(\eta)}{w'(\eta)} d\eta = 0 \tag{6.5.21}$$

$$\int_{\Gamma} \frac{\phi(\eta^{-1})}{\eta-\zeta} d\eta = 0 \tag{6.5.22}$$

In closed contour Γ we have $L_1 = -L_6$, $L_2 = -L_5$ and $L_3 = -L_4$, therefore, $\Gamma = \gamma + \Gamma_1$ and Eqs.(6.5.20), (6.5.21) and (6.5.22) will change to:

$$\int_{\Gamma_1+\gamma} \frac{\phi(\eta^{-1})}{\eta-\zeta} d\eta = 0 \Rightarrow \int_{\gamma} \frac{\phi(\eta^{-1})}{\eta-\zeta} d\eta = -\int_{\Gamma_1} \frac{\phi(t^{-1})}{t-\zeta} dt \tag{6.5.23}$$

$$\int_{\Gamma_1+\gamma} \frac{w(\eta^{-1})}{\eta-\zeta} \frac{\phi'(\eta)}{w'(\eta)} d\eta = 0 \Rightarrow \int_{\gamma} \frac{w(\eta^{-1})}{\eta-\zeta} \frac{\phi'(\eta)}{w'(\eta)} d\eta = -\int_{\Gamma_1} \frac{w(t^{-1})}{t-\zeta} \frac{\phi'(t)}{w'(t)} dt \tag{6.5.24}$$

$$\int_{\Gamma_1+\gamma} \frac{\psi_I(\eta)}{\eta-\zeta} d\eta = 0 \Rightarrow \int_{\gamma} \frac{\psi_I(\eta)}{\eta-\zeta} d\eta = -\int_{\Gamma_1} \frac{\psi_I(t)}{t-\zeta} dt \tag{6.5.25}$$

By substituting Eqs. (6.5.23), (6.5.24) and (6.5.25) into (6.5.19) we have:

$$\int_{\Gamma_1} \frac{\psi_I(t)}{t-\zeta} dt + \int_{\Gamma_1} \frac{w(t^{-1})}{t-\zeta} \frac{\phi'(t)}{w'(t)} dt + \int_{\Gamma_1} \frac{\phi(t^{-1})}{t-\zeta} dt = 0 \tag{6.5.26}$$

In the integrals (6.5.26), inside the contour Γ_1 the only singularity is $t=\zeta$. Therefore, each integral can be evaluated by using the residue formula, that is, $\int_{\Gamma_1}\frac{f(z)}{z-a}dz=2\pi i\,f(a)$ then:

$$2\pi i\left(\psi_I(\zeta)+w(\zeta^{-1})\frac{\phi'(\zeta)}{w'(\zeta)}+\phi(\zeta^{-1})\right)=0 \tag{6.5.27}$$

From Eq. (6.5.27) we can find the relationship between $\psi_I(\zeta)$ versus $\phi(\zeta)$ as follows:

$$\psi_I(\zeta)=-w\left(1/\zeta\right)\frac{\phi'(\zeta)}{w'(\zeta)}-\phi\left(1/\zeta\right) \tag{6.5.28}$$

This is known as Muskhlishivili formulas relating $\phi(\zeta)$ and $\psi(\zeta)$, and we can use it to find out the stress formulas around an elliptical hole in the next article.

6.6 FUNCTIONS $\phi(\zeta)$ AND $\psi(\zeta)$ AROUND ELLIPTIC HOLE

In fracture mechanics, usually the face of the crack is not loaded. Therefore, determination of $\phi(\zeta)$ and $\psi(\zeta)$ is easier. In this section we assume an elliptical hole which unloaded and therefore, we can use Eq. (6.5.28). Moreover, $\phi(z)$ and $\psi(z)$ consists of two parts. The first part is discussed in Eq. (6.5.9), and will be written again: $\phi(z)\underset{z\to\infty}{=}\frac{\sigma}{4}z\Rightarrow\phi(\zeta)\underset{\zeta\to\infty}{=}\frac{\sigma}{4}\underbrace{\left(\zeta+\frac{m}{\zeta}\right)c}_{z}$ which can be simplified to:

$$\phi(\zeta)\underset{\zeta\to\infty}{=}\frac{1}{4}\sigma c\zeta \tag{6.6.1}$$

The 2nd part of $\phi(\zeta)$ that will combined with (1) is a result of elliptical hole so that:

$$\phi(\zeta)=\frac{1}{4}\sigma c\zeta+\sum_{n=1}^{\infty}A_n\zeta^{1-2n} \tag{6.6.2}$$

As we can see in Eq. (6.6.2) the power of ζ is always negative because it should be vanish in the distance far from the hole, that is, $\lim_{\zeta\to\infty} \sum_{n=1}^{\infty} A_n \zeta^{1-2n} = 0$. Moreover, the power of ζ is odd since from Eq. (6.6.2) we can have:

$$\phi(-\zeta) = -\frac{1}{4}\sigma c\zeta - \sum_{n=1}^{\infty} A_n \zeta^{1-2n} = -\phi(\zeta) \tag{6.6.3}$$

The Eq. (6.6.3) shows that $\phi(-\zeta) = -\phi(\zeta)$, which is called central symmetric condition, whereas for $\sum_{n=1}^{\infty} A_n \zeta^{-2n}$ we cannot have this condition and $\phi(-\zeta) \neq -\phi(\zeta)$. The next issue is determination of the coefficients A_n, and this can be achieved by using 6.5.28 to find $\psi_I(\zeta)$. Also $\psi_I(\zeta)$ cannot have power of ζ above 1. In Eq. (6.5.28) there are several terms that each term should be analyzed individually such as:

$$w(\zeta) = \left(\zeta + \frac{m}{\zeta}\right)c \Rightarrow w\left(\frac{1}{\zeta}\right) = \left(m\zeta + \frac{1}{\zeta}\right)c \tag{6.6.4}$$

$$w'(\zeta) = \left(1 - \frac{m}{\zeta^2}\right)c \tag{6.6.5}$$

By dividing Eqs. (6.6.4) to (6.6.5) we have:

$$\frac{w\left(\frac{1}{\zeta}\right)}{w'(\zeta)} = \frac{\left(m\zeta + \frac{1}{\zeta}\right)c}{\left(1 - \frac{m}{\zeta^2}\right)c} = \frac{m\zeta^2 + 1}{\zeta^2 - m}\zeta \tag{6.6.6}$$

$\phi(\zeta) = \frac{1}{4}\sigma c\zeta + \sum_{n=1}^{\infty} A_n \zeta^{1-2n} \Rightarrow \phi'(\zeta) = \frac{1}{4}\sigma c + \sum_{n=1}^{\infty} A_n \zeta^{-2n}$ which can be written into:

$$-\phi'(\zeta) = -\frac{1}{4}\sigma c - \zeta^{-1} \sum_{n=1}^{\infty} A_n \zeta^{1-2n} \tag{6.6.7}$$

This Eq. (6.6.6) enables finding out

$$-w\left(\frac{1}{\zeta}\right)\frac{\phi'(\zeta)}{w'(\zeta)}=\frac{m\zeta^2+1}{\zeta^2-m}\zeta\left(-\frac{1}{4}\sigma c-\zeta^{-1}\sum_{n=1}^{\infty}A_n\zeta^{1-2n}\right)\text{ or:}$$

$$-w\left(\frac{1}{\zeta}\right)\frac{\phi'(\zeta)}{w'(\zeta)}=-\frac{m\zeta^2+1}{\zeta^2-m}\left(\frac{1}{4}\sigma c\zeta+\sum_{n=1}^{\infty}A_n\zeta^{1-2n}\right) \tag{6.6.8}$$

The other term in Eq. (6.5.28) can be found like this:

$$\phi(\zeta)=\frac{1}{4}\sigma c\zeta+\sum_{n=1}^{\infty}A_n\zeta^{1-2n}\Rightarrow-\phi\left(\frac{1}{\zeta}\right)=-\frac{\sigma c}{4\zeta}-\sum_{n=1}^{\infty}A_n\zeta^{2n-1} \tag{6.6.9}$$

Now all the components of Eq. (6.5.28) or $\psi_I(\zeta)=-w\left(\frac{1}{\zeta}\right)\frac{\phi'(\zeta)}{w'(\zeta)}-\phi\left(\frac{1}{\zeta}\right)$ are known, that is,

$$\psi_I(\zeta)=-\frac{m\zeta^2+1}{\zeta^2-m}\left(-\frac{1}{4}\sigma c\zeta-\sum_{n=1}^{\infty}A_n\zeta^{1-2n}\right)-\frac{\sigma c}{4\zeta}-\sum_{n=1}^{\infty}A_n\zeta^{2n-1} \tag{6.6.10}$$

We discussed that $\psi_I(\zeta)$ cannot have power of ζ above 1. therefore we need the term $n=1$

$$\psi_I(\zeta)=-\frac{m\zeta^2+1}{\zeta^2-m}\left(\frac{\sigma c\zeta}{4}+A_1\zeta^{-1}\right)-\frac{\sigma c}{4\zeta}-A_1\zeta \tag{6.6.11}$$

Now in Eq. (6.6.11) we set $\zeta\to\infty$ then we have $\lim_{\zeta\to\infty}\frac{m\zeta^2+1}{\zeta^2-m}=m$ and Eq. (6.6.11) changes to:

$$\lim_{\zeta\to\infty}\psi_I(\zeta)=-\frac{m\sigma c}{4}\zeta-A_1\zeta \tag{6.6.12}$$

Combining the Eq. (6.5.10), with the definition (6.5.16) provides a new form:

$$\psi_I(z)_{z\to\infty}=\frac{\sigma}{2}z \tag{6.6.13}$$

In w plane the Eq. (6.6.13) will change to:

$$\underset{\zeta\to\infty}{\psi_I(\zeta)}=\frac{\sigma}{2}\left(\zeta+\frac{m}{\zeta}\right)c \Rightarrow \underset{\zeta\to\infty}{\psi_I(\zeta)}=\frac{\sigma c \zeta}{2} \tag{6.6.14}$$

By comparing Eqs. (6.6.14) and (6.6.12) we can find that:

$$-\frac{m\sigma c}{4}-A_1=\frac{\sigma c}{2} \Rightarrow A_1=-\frac{\sigma c}{4}(2+m) \tag{6.6.15}$$

Now we substitute Eq. (6.6.15) into Eq. (6.6.12) and considering that other coefficients are zero (see Eq. (6.6.11)), the expression for $\phi(\zeta)$ will be:

$$\phi(\zeta)=\frac{\sigma c}{4}\left(\zeta-(2+m)\zeta^{-1}\right) \tag{6.6.16}$$

Equation (6.6.16) is very important and it will be used later. It is also better to find the expression for $\psi_I(\zeta)$ as well. First we need to calculate the following:

$$\phi\left(\frac{1}{\zeta}\right)=\frac{\sigma c}{4}\left(\zeta^{-1}-(2+m)\zeta\right) \tag{6.6.17}$$

$$\phi'(\zeta)=\frac{\sigma c}{4}\left(1+(2+m)\zeta^{-2}\right) \tag{6.6.18}$$

Then by substituting Eqs. (6.6.6), (6.6.17) and (6.6.18) into Eq. (6.5.28) we can find $\psi_I(\zeta)$.

$$\psi_I(\zeta)=\frac{\sigma c}{4}\left[-\left(1+(2+m)\zeta^{-2}\right)\frac{m\zeta^2+1}{\zeta^2-m}\zeta-\frac{1-(2+m)\zeta^2}{\zeta}\right]$$

Exercise 6.6.1: Simplify the above expression

Solution: $\psi_I(\zeta)=\frac{\sigma c}{4}\left[\frac{-(\zeta^2+2+m)(m\zeta^2+1)}{\zeta(\zeta^2-m)}-\frac{1-(2+m)\zeta^2}{\zeta}\right]$ further simplification can be done by considering the common denominator, that is,

$$\psi_I(\zeta)=\frac{\sigma c}{4}\left[\begin{matrix}-(\zeta^2+2+m)(m\zeta^2+1)-\\(\zeta^2-m)(1-(2+m)\zeta^2)\end{matrix}\right]\frac{1}{\zeta(\zeta^2-m)}$$ followed by:

$$\psi_I(\zeta)=\frac{\sigma c}{4}\left[\begin{matrix}\left(\overbrace{2-m+m}^{2}\right)\zeta^4+\left(\overbrace{-1-2m-m^2-1-m^2-2m}^{-2(1+2m+m^2)}\right)\\ \zeta^2+(-2-m+m)\end{matrix}\right]\frac{1}{\zeta(\zeta^2-m)}$$

or:

$$\psi_I(\zeta)=\frac{\sigma c}{2}\left[\frac{\zeta^4-(1+2m+m^2)\zeta^2-1}{\zeta(\zeta^2-m)}\right] \tag{6.6.19}$$

Equation (6.6.19) will be used later and the exercise shows how the lengthy algebraic manipulations can be for finding a formula. Now we describe σ_n and σ_t which are the stresses at normal and tangential directions to the elliptical hole faces at region near the hole. It is obvious that σ_x and σ_y should be rotated by a particular angle to become σ_n and σ_t but under this rotation the sum $\sigma_x+\sigma_y$ and $\sigma_n+\sigma_t$ does not change. This can be observed by Mohr circle in elementary solid mechanics. Now Eq. (6.3.9) can be written again into this form:

$$\sigma_x+\sigma_y=\sigma_n+\sigma_t=4\Re\left(\phi'(z)\right) \tag{6.6.20}$$

Moreover, $\phi'(z)=\frac{d\phi}{dz}=\frac{d\phi}{d\zeta}\frac{d\zeta}{dz}=\frac{d\phi}{d\zeta}\left(\frac{dz}{d\zeta}\right)^{-1}$, and since $\frac{dz}{d\zeta}=w'(\zeta)$ we can write:

$$\phi'(z)=\frac{\phi'(\zeta)}{w'(\zeta)} \tag{6.6.21}$$

Comparing (6.6.20) and (6.6.21) gives the following:

$$\sigma_n+\sigma_t=4\Re\left(\frac{\phi'(\zeta)}{w'(\zeta)}\right) \tag{6.6.22}$$

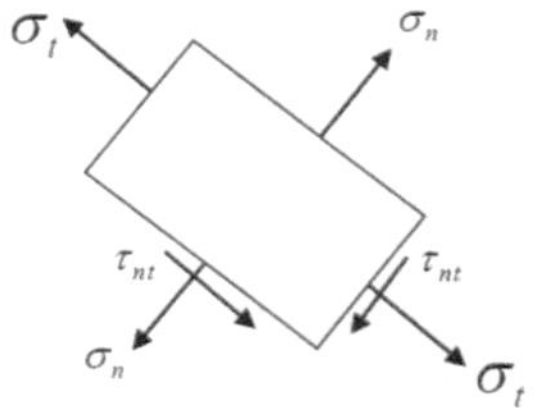

FIGURE 6.6.1 σ_n and σ_t

Equation (6.6.22) is expressed in w plane and on the elliptical hole (boundary) there is not normal stress $\sigma_n = 0$ and only σ_t exists. We discussed that hole or crack itself in w plane can be shown by a unit circle or by $\zeta = \eta$. Therefore, Eq. (6.6.22) can be expressed in terms of η valid only on the hole, that is,

$$(\sigma_n + \sigma_t)_{\zeta=\eta} = 4\Re\left(\frac{\phi'(\eta)}{w'(\eta)}\right) \tag{6.6.23}$$

If we substitute Eqs. (6.6.5) and (6.6.18) into Eq. (6.6.22) we have:

$$\sigma_n + \sigma_t = 4\Re\left(\frac{\frac{\sigma c}{4}\left(1+(2+m)\zeta^{-2}\right)}{\left(1-\frac{m}{\zeta^2}\right)c}\right) \text{ simple form:}$$

$$\sigma_n + \sigma_t = \sigma\,\Re\left(\frac{\zeta^2+2+m}{\zeta^2-m}\right) \tag{6.6.24}$$

On the elliptical hole Eq. (6.6.24) will be:

$$(\sigma_n + \sigma_t)_{\zeta=\eta} = 4\Re\left(\frac{\eta^2+2+m}{\eta^2-m}\right) \tag{6.6.25}$$

Exercise 6.6.2: Over the hole $\sigma_n = 0$, determine σ_t versus σ

Solution: In (6.6.25) we set $\eta = e^{i\theta}$ we see that $\frac{\sigma_t}{\sigma} = \Re\left(\frac{e^{2i\theta}+2+m}{e^{2i\theta}-m}\right)$ or

$$\frac{\sigma_t}{\sigma} = \Re\left(\frac{\left(e^{2i\theta}+2+m\right)\left(e^{-2i\theta}-m\right)}{\left(e^{2i\theta}-m\right)\left(e^{-2i\theta}-m\right)}\right) \text{ results}$$

$$\frac{\sigma_t}{\sigma} = \Re\left(\frac{-m^2 - 2m - m e^{2i\theta} - (m+2)e^{-2i\theta} + 1}{m^2 + 1 - m\left(e^{2i\theta} + e^{-2i\theta}\right)} \right)$$

$e^{2i\theta} + e^{-2i\theta} = 2\cos 2\theta$ and $e^{\pm 2i\theta} = \cos 2\theta \pm i \sin 2\theta$ by substituting and separating the real part, we can write:

$$\frac{\sigma_t}{\sigma} = \frac{1 - m^2 - 2m + 2\cos 2\theta}{m^2 + 1 - 2m\cos 2\theta} \tag{6.6.26}$$

The maximum stress occurs at $\theta = 0$ i.e.

$$\frac{\sigma_{\max}}{\sigma} = \frac{3 - m^2 - 2m}{m^2 + 1 - 2m} = \frac{3+m}{1-m} \tag{6.6.27}$$

From 6.5.11 we substitute m in (6.6.27) then we have:

$$\sigma_{\max} = \left(1 + \frac{2a}{b}\right)\sigma \tag{6.6.28}$$

A crack can be expressed by an ellipse in which $a >> b$ and implies that $\frac{2a}{b} >> 1$ and from Eq. (6.6.28) it can be concluded that for a crack $\sigma_{\max} \to \infty$ regardless of how big or small σ is. This means that in presence of cracks stress theories does not work and we need to suggest other theories, which fracture mechanics, are based upon. We will study those theories in the next chapters.

KEYWORDS

- **airy stress function**
- **bi-harmonic equation**
- **Cauchy-Riemann condition**
- **harmonic functions**
- **linear elasticity**

CHAPTER 7

DERIVATION OF FRACTURE MECHANICS FROM LINEAR ELASTICITY

CONTENTS

7.1 FUNCTIONS $\phi(z)$ AND $\psi(z)$ FOR A CRACK: DETERMINATION OF STRAIN ENERGY

A crack can be assumed as an ellipse in which $c=\frac{a+b}{2}=\frac{a}{2}$, and eccentricity $m=\frac{a-b}{a+b}=1$. Therefore, Eq. (6.6.16) can be written into this form:

$$\phi(\zeta)=\frac{\sigma a}{8}\left(\zeta-(2+1)\zeta^{-1}\right)$$

Simplified to:

$$\phi(\zeta)=\frac{\sigma a}{8}\left(\zeta-3/\zeta\right) \qquad (7.1.1)$$

Moreover, we substitute $c=\frac{a}{2}$, and $m=1$, into 6.6.19 then we have:

$$\psi_I(\zeta)=\frac{\sigma a}{4}\left[\frac{\zeta^4-(1+2+1)\zeta^2-1}{\zeta(\zeta^2-1)}\right]$$

and can be simplified:

$$\psi_I(\zeta)=\frac{\sigma a}{4}\left[\frac{\zeta^3-4\zeta-\zeta^{-1}}{\zeta^2-1}\right] \tag{7.1.2}$$

The mapping function of a crack into outside of a unit circle can found by substituting $c=\frac{a}{2}$ and $m=1$ into 6.5.14, that is,

$$z=\frac{a}{2}\left(\zeta+\frac{1}{\zeta}\right) \tag{7.1.3}$$

Exercise 7.1.1: for a crack find ζ versus z

Solution: by squaring both sides of Eq. (7.1.3) we have, $z^2=\frac{a^2}{4}\left(\zeta^2+\frac{1}{\zeta^2}+2\right)$, and then by deducting a^2 from both sides we have: $z^2-a^2=\frac{a^2}{4}\left(\zeta^2+\frac{1}{\zeta^2}-2\right)=\frac{a^2}{4}\left(\zeta-\frac{1}{\zeta}\right)^2$ and taking square root of this provides:

$$\sqrt{z^2-a^2}=\frac{a}{2}\left(\zeta-\frac{1}{\zeta}\right) \tag{7.1.4}$$

Then from Eqs. (7.1.3) and (7.1.4) ζ can be found versus *z*, that is,

$$\zeta=\frac{\sqrt{z^2-a^2}+z}{a} \qquad \frac{1}{\zeta}=\frac{z-\sqrt{z^2-a^2}}{a} \tag{7.1.5}$$

By substituting Eqs. 7.1.5) into (7.1.1) we can change $\phi(\zeta)$ into $\phi(z)$, that is,

$$\phi(z)=\frac{\sigma a}{8}\left(\frac{\sqrt{z^2-a^2}+z}{a}-\frac{3z-3\sqrt{z^2-a^2}}{a}\right)$$

which can be simplified to:

$$\phi(z)=\frac{\sigma}{4}\left(2\sqrt{z^2-a^2}-z\right) \tag{7.1.6}$$

Exercise 7.1.2: Convert $\psi_I(\zeta)$ into $\psi_I(z)$

Solution: Similarly we substitute Eq. (7.1.5) into Eq. (7.1.2) and we have:

$$\psi_I(z)=\frac{\sigma a}{4}\left[\frac{\frac{1}{a^3}\left(\sqrt{z^2-a^2}+z\right)^3-\frac{4}{a}\left(\sqrt{z^2-a^2}+z\right)-\frac{1}{a}\left(z-\sqrt{z^2-a^2}+z\right)}{\frac{1}{a^2}\left(\sqrt{z^2-a^2}+z\right)^2-1}\right]$$

and can be simplified to,

$$\psi_I(z)=\frac{\sigma a}{4}\left[\frac{\frac{1}{a}\left(\sqrt{z^2-a^2}+z\right)^3-4a\left(\sqrt{z^2-a^2}+z\right)-a\left(z-\sqrt{z^2-a^2}\right)}{2\sqrt{z^2-a^2}\left(\sqrt{z^2-a^2}+z\right)}\right]$$

Further simplification $\psi_I(z)=\frac{\sigma}{8\sqrt{z^2-a^2}}\left[\left(\sqrt{z^2-a^2}+z\right)^2-4a^2-a^2\frac{z-\sqrt{z^2-a^2}}{\sqrt{z^2-a^2}+z}\right]$ this gives, $\psi_I(z)=\frac{\sigma}{8\sqrt{z^2-a^2}}\left[\left(\sqrt{z^2-a^2}+z\right)^2-4a^2-\left(z-\sqrt{z^2-a^2}\right)^2\right]$ the expression inside square bracket can be simplified as well, $\psi_I(z)=\frac{\sigma}{8\sqrt{z^2-a^2}}\left[4z\sqrt{z^2-a^2}-4a^2\right]$ or finally:

$$\psi_I(z) = \left(z - \frac{a^2}{\sqrt{z^2 - a^2}} \right) \frac{\sigma}{2} \tag{7.1.7}$$

Now that we have Eqs. (7.1.6) and (7.1.7), we can rewrite Eq. (6.4.27) in terms $\psi_I(z)$:

$$2\mu(u + iv) = \kappa\phi(z) - z\overline{\phi'(z)} - \overline{\psi_I(z)} \tag{7.1.8}$$

Differentiation of Eq. (7.1.6) versus z provides $\phi'(z)$, that is,

$$\phi'(z) = \frac{\sigma}{4}\left(\frac{2z}{\sqrt{z^2 - a^2}} - 1 \right) = \frac{\sigma}{2}\left(\frac{z}{\sqrt{z^2 - a^2}} - \frac{1}{2} \right) \tag{7.1.9}$$

$$\overline{\phi'(z)} = \frac{\sigma}{2}\overline{\left(\frac{z}{\sqrt{z^2 - a^2}} - \frac{1}{2} \right)} \tag{7.1.10}$$

From Eq. (7.1.7) we need to find:

$$\overline{\psi_I(z)} = \overline{\left(z - \frac{a^2}{\sqrt{z^2 - a^2}} \right)} \frac{\sigma}{2} \tag{7.1.11}$$

The objective is finding the displacement v of the face of the crack. It is obvious that we need to change z into x and the crack region can be designated by $-a \le x \le a$.

Exercise 7.1.3: Determine all the terms of Eq. (7.1.8) on the track expressed by $-a \le x \le a$

Solution: It should be remembered that in region $-a \le x \le a$ we have, $\sqrt{x^2 - a^2} = i\sqrt{a^2 - x^2}$ and the individual terms will be:
$\kappa\phi(x) = \kappa\frac{\sigma}{4}\left(2\sqrt{x^2 - a^2} - x\right) = \kappa\frac{\sigma}{2}\left(i\sqrt{a^2 - x^2} - 0.5x\right)$ the other term is:

$\phi^{'}(x)=\frac{\sigma}{2}\left(\frac{x}{\sqrt{x^2-a^2}}-\frac{1}{2}\right)=\frac{\sigma}{2}\left(\frac{-i\,x}{\sqrt{a^2-x^2}}-\frac{1}{2}\right)$ and therefore $\overline{x\phi^{'}(x)}$ can be written by:

$-x\overline{\phi^{'}(x)}=-\frac{\sigma}{2}x\left(\frac{i\,x}{\sqrt{a^2-x^2}}-\frac{1}{2}\right)$, the last term that needs to be found is $\overline{\psi_I(x)}$ as follows:

$$\psi_I(x)=\left(x-\frac{a^2}{\sqrt{x^2-a^2}}\right)\frac{\sigma}{2}=\left(x+\frac{a^2 i}{\sqrt{a^2-x^2}}\right)\frac{\sigma}{2} \text{ resulting}$$

$$\overline{\psi_I(x)}=\left(x-\frac{a^2 i}{\sqrt{a^2-x^2}}\right)\frac{\sigma}{2}$$

Now we collect all the imaginary parts of the components in (7.1.8) and equate them with imaginary part of the left hand side of (7.1.8) then we have:

$2\mu\, v_{z=x} i=\kappa\frac{\sigma}{2}\left(i\sqrt{a^2-x^2}\right)-\frac{\sigma}{2}\left(\frac{i\,x^2}{\sqrt{a^2-x^2}}\right)+\left(\frac{a^2 i}{\sqrt{a^2-x^2}}\right)\frac{\sigma}{2}$ which easily results:

$$2\mu\, v=(\kappa+1)\frac{\sigma}{2}\sqrt{a^2-x^2}$$

The above expression can be rewritten into the following form:

$$v(x)=\left(\frac{\kappa+1}{4\mu}\right)\sigma\sqrt{a^2-x^2} \qquad (7.1.12)$$

Equation (7.1.12) gives the normal displacement v of the crack as a result of remote stress σ, it indicates that if σ increases the normal displacement v also increases proportional, thereby increasing the strain energy.

In the last chapter in Eq. (6.6.28), we saw that even a small remote stress σ can cause an infinite stress ($\sigma_{\max}\rightarrow\infty$), this Eq. (6.6.28) was also demonstrated previously by Inglis in 1913, using a different approach and it could not solely answer the mysteries about fracture. However,

Eq. (7.1.12) demonstrates that small σ may produce infinite stress but it produces small v, and consequently cannot produce enough stain energy, to tear off the object through the crack.

Later Irwin called this type of fracture mode I and named them tearing mode. Scientists and mathematicians before Irwin hardy tried to come up, with alternative criteria for describing fracture by cracks. In 1921 Griffith gave the first theory, simply relying on Inglis formulas and approach. However, he tried to find a formula for the acquired stain energy in presence of a crack he did not do it via Eq. (7.1.12) but used a formidable procedure to find out and this enabled him to develop the first theory for fracture. Herein we develop his formula using our approach.

Figure 7.2, demonstrates that if we want to transform a cracked plate into a crackles plate we need to close the crack by compressive stress σ and we need to apply negative work against the crack. The extra

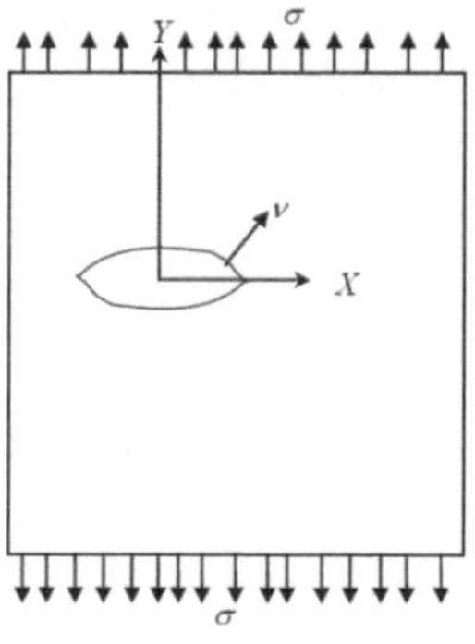

FIGURE 7.1 Crack displacement.

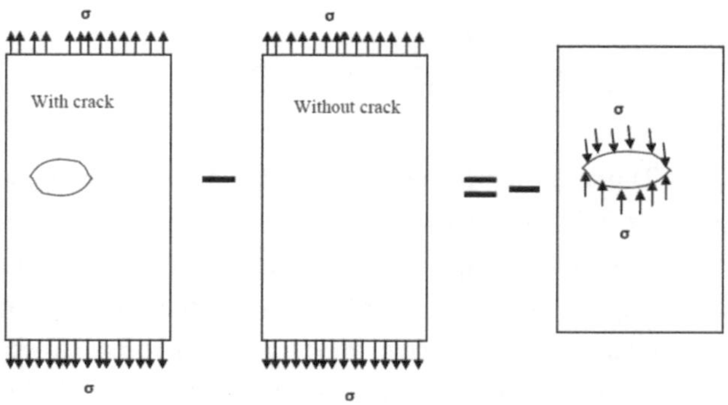

FIGURE 7.2 Superposition principle in describing extra strain energy acquired by a crack.

strain energy is just the same work but with positive sign, therefore we can write: $U = \left[-\frac{1}{2} B \int_{-a}^{a} -\sigma\, v(x)\, dx \right]$ in which B is the plate thickness so that $B\,dx = dA$ and $dF = \sigma\, dA$, therefore, $U = \left[\frac{1}{2} \int_{-a}^{a} \underbrace{\sigma\, B\, dx}_{dF}\, v(x) \right] = \frac{B}{2} \int_{-a}^{a} \sigma\, v(x)\, dx$ and by substituting (7.1.12) for $v(x)$, we have: $U = \frac{\sigma B}{2} \int_{-a}^{a} \left(\frac{\kappa + 1}{4\mu} \right) \sigma \sqrt{a^2 - x^2}\, dx$ which results:

$$U = \left(\frac{\kappa + 1}{4\mu} \right) \frac{\sigma^2 B}{2} \int_{-a}^{a} \sqrt{a^2 - x^2}\, dx \qquad (7.1.13)$$

To find the integral $\int_{-a}^{a} \sqrt{a^2 - x^2}\, dx$ we set $x = a\cos\theta \Rightarrow dx = -a\sin\theta\, d\theta$, and therefore we have: $\int_{-a}^{a} \sqrt{a^2 - x^2}\, dx = \int_{\pi}^{0} \sqrt{a^2 - a^2\cos^2\theta}\, (-a\sin\theta\, d\theta) = \frac{\pi a^2}{2}$ and this integral is for one face of the crack, for two faces $2\int_{-a}^{a} \sqrt{a^2 - x^2}\, dx = \pi a^2$ and therefore the strain energy in Eq. (7.1.13) will change to:

$$U = \left(\frac{\kappa + 1}{8\mu} \right) B\pi a^2 \sigma^2 \qquad (7.1.14)$$

If we want to express the strain energy per unit thickness it will be:

$$U = \left(\frac{\kappa + 1}{8\mu} \right) \pi a^2 \sigma^2 \qquad (7.1.15)$$

Using the formula in Eq. (7.1.15) in the next section, we can describe the fracture theory described by Griffith in 1921. The theory that completed later by other scientists.

7.2 GRIFFITH THEORY FOR BRITTLE FRACTURE

The way Griffith looks at a crack is, seeing rest of the object glued together (except crack faces). He says a surface resistance R is kept the faces of crackles parts together. Fig. 7.3, shows R concept. Its unit is different from stress, that is, J/m^2 or energy per unit surface that becomes F/L. He is saying for tearing off a surface and extending a crack we need to do mechanical work. This work can be done by the force P in Fig. 7.3. This work will be done against the resistance R and the ΔW_c is defined by:

$$\Delta W_c = R\,B\,\Delta(2a) \tag{7.2.1}$$

Figure 7.3 and Eq. (7.2.1) confirms that for growing the crack length by Δa, the required work is ΔW_c, and some material like glass he arranged experiments and calculated R. Further form of Eq. (7.2.1) is written like:

$$\frac{dW_c}{da} = \underbrace{R}_{resistance}\ \overbrace{B}^{thickness} \tag{7.2.2}$$

In previous section we discussed that in presence of a crack extra strain energy is gained that was given by Eq. (7.1.14), that is,

$$U = \left(\frac{\kappa+1}{8\mu}\right) B\pi\, a^2\sigma^2$$

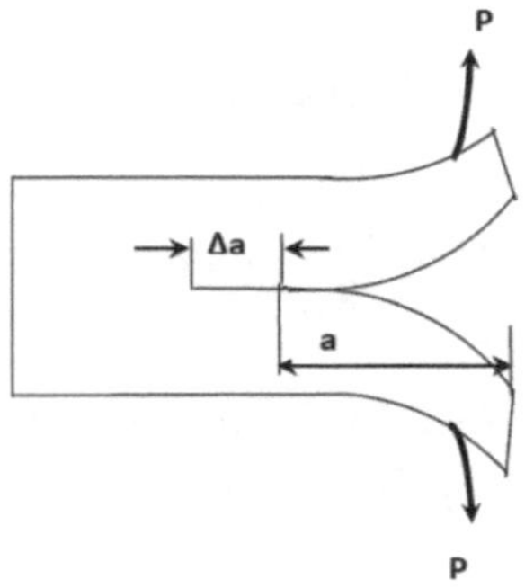

FIGURE 7.3 R concept.

Assuming that the crack growth is $\Delta\ a$ and the extra strain energy will also grow by $\Delta\ U$, so that:

$$\Delta U = \left(\frac{\kappa+1}{8\mu}\right) B\pi\, a\sigma^2 \Delta(2a) \tag{7.2.3}$$

The above equation can be written into this form as well:

$$\frac{dU}{d(2a)} = \left(\frac{\kappa+1}{\mu}\right)\frac{\pi\,\sigma^2}{8} a\,B \tag{7.2.4}$$

Now we draw the functions:

$$U = \left(\frac{\kappa+1}{8\mu}\right)\pi\, a^2\sigma^2 \tag{7.2.5}$$

And also:

$$W_c = -2\,R\,a \Rightarrow |W_c| = 2\,R\,a \tag{7.2.6}$$

Together in Fig. 7.4, we see that, U versus a is a parabolic curve while $|W_c|$ versus a is a straight line passing though the origin. If the stress σ and the crack length a is such that energy $U < |W_c|$, then there is not enough energy to tear off the surface and therefore, the crack will not grow. However, if we have $U > |W_c|$ the crack will grow, since there is enough energy to tear off the surface. The critical point is when the straight line and parabola in Fig. 7.4, intersect so that $U = |W_c|$. According to Eq. (7.2.6) we defined W_c as a negative value and the total work W is:

$$W = U + W_c \tag{7.2.7}$$

The critical point means that sum of the total work will be zero, and the external stresses when a is above a particular value can produce enough energy to overcome W_c. To guarantee that crack growing continues, we need to analyze differentiated form of Eq. (7.2.7), that is,

$\frac{dW}{da} = \frac{dU}{da} + \frac{dW_c}{da}$, then we can say that at point $\frac{dW}{da} = 0$, fast fracture occurs, and according to Fig. 7.4, we can have: $\frac{dW}{da} \geq 0 \Rightarrow \frac{dU}{da} \geq -\frac{dW_c}{da}$ and if we substitute from Eqs. (7.2.5) and (7.2.6) we have:

$$\left(\frac{\kappa+1}{8\mu}\right)\pi a\sigma^2 \geq R \tag{7.2.8}$$

It should be reminded that $\frac{dU}{da}$ is an important quantity in fracture mechanics, named "energy release rate", it is shown by G (not shear modulus) it value which is taken from inequality Eq. (7.2.8) and is:

$$G = \left(\frac{\kappa+1}{8\mu}\right)\pi a\sigma^2 \tag{7.2.9}$$

When we substitute Eq. 7.2.9) into the inequality Eq. (7.2.8) it becomes: $G_c \geq R$, the question is, why we replace G with G_c? The answer explains basics of the Griffith theory, saying that if the applied stress σ and also the crack length a, makes G in Eq. (7.2.9) such big that reaches a onset of G_c, then Eq. (7.2.8) can be satisfied and fast fracture will occur in this situation $G_c = R$, in many books the formula for U is expressed without B, they are also equally correct since they have assumed a unit thickness or $B = 1$. Throughout this book we have used both forms, when it is necessary.

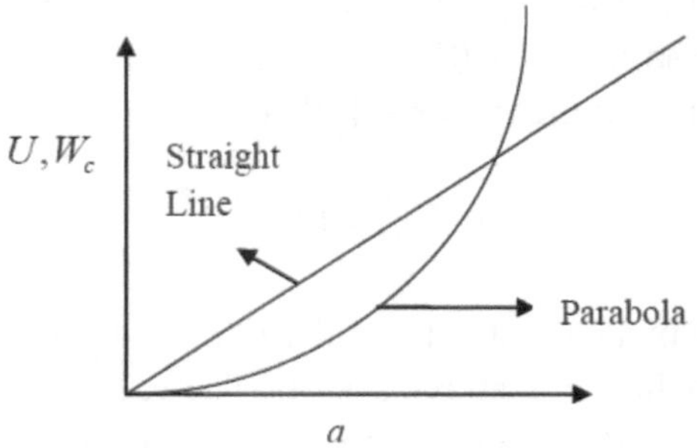

FIGURE 7.4 *U and* $|W_c|$.

Equations (7.2.8) and (7.2.9) explains Griffith theory completely. Later Irwin expressed the Griffith theory with different language in order to explain we go to Eq. (6.4.27) and for plain stress case:

$$\kappa = \frac{3-\upsilon}{1+\upsilon} \tag{7.2.10}$$

$$\mu = \frac{E}{2(1+\upsilon)} \tag{7.2.11}$$

Then $\kappa + 1 = \frac{3-\upsilon}{1+\upsilon} + 1 = \frac{4}{1+\upsilon}$ which results $\frac{\kappa+1}{\mu} = \frac{4}{1+\upsilon}\frac{2(1+\upsilon)}{E} = \frac{8}{E}$, by substituting these into Eq. (7.2.9) we have:

$$G = \left(\frac{8}{8E}\right)\pi a\sigma^2 \Rightarrow G = \frac{\pi a\sigma^2}{E} \tag{7.2.12}$$

Based on Eq. (7.2.12) we define a new parameter:

$$K_I = \sigma\sqrt{\pi a} \tag{7.2.13}$$

We call it Stress Intensity Factor (SIF) in mode I (tearing mode). K_I depends on both applied stress σ and also crack length a, by increasing each of them K_I also increases and another form of Eq. (7.2.12) that written in terms of K_I, can be found by substituting Eq. (7.2.12) into Eq. (7.2.13).

$$K_I = \sqrt{E\,G} \tag{7.2.14}$$

In Eq. (7.2.14) E is the elastic modulus and is a constant value. If K_I increases it effects G and it increases as well. If K_I reaches a critical value K_{IC}, that is, ($K_I \to K_{IC}$) so that it makes $G \to G_C$, then Eq. (7.2.14) will change to its new form written by:

$$K_{IC} = \sqrt{E\,G_C} \tag{7.2.15}$$

In Eq. (7.2.15) K_{IC} is critical stress intensity factor and a very important quantity in fracture mechanics so that Irwin gave it a new name called "Fracture Toughness." Its unit is unusual and according to Eq. (7.2.13) should be Mpa.m$^{1/2}$ (different from stress). Equation (7.2.15) can be assumed heart of fracture mechanics. It states that if the applied stress and existing crack length in an object is such that $K_I \rightarrow K_{IC}$, then definitely $G \rightarrow G_C$ and since $G_c \geq R$ the crack length increases nonstop without control. We call this situation "fast fracture."

K_{IC} is very important in fracture mechanics, and all the attempts are concentrated on calculating K_{IC}, either by experimental, or analytical or numerical methods. Some material like glass has low K_{IC} which means they break soon they are also called brittle materials. In these materials all the deformations and stresses remain in elastic region. The science by which we study their fracture is called Linear Elastic Fracture Mechanic, abbreviated by LEFM.

Other types of materials like steel have high K_{IC}, which means they do not break easily and they are called "ductile materials." In these materials there are significant plastic deformation around the crack tip and they cannot be analyzed directly by LEFM. Alternatively there is an established field named Elastic Plastic Fracture Mechanics (EPFM) that can be used to analyze the fracture of the ductile materials. For ductile materials not only K_{IC} is important the G_c is also equally important and it can be found by experimental methods and also numerical methods. It can be shown in future chapters that one of the powerful numerical methods, for calculating G_c is called J integral. In this method we can simply replace G_c by a new parameter J, that expresses the J integral.

Whatever mentioned by now, is for the plane stress case. For plane strain case we need to refer to 6.4.27 and we can see that, $\kappa = 3 - 4\upsilon$ and $\kappa + 1 = 3 - 4\upsilon + 1 = 4(1-\upsilon)$, therefore we have $\frac{\kappa+1}{\mu} = 4(1-\upsilon)\frac{2(1+\upsilon)}{E} = \frac{8(1-\upsilon^2)}{E}$ and if we substitute this into (7.2.9) we have: $G = \left(\frac{\kappa+1}{8\mu}\right)\pi a\sigma^2 \Rightarrow G = \frac{8(1-\upsilon^2)}{8E}\pi a\sigma^2$ which can be simplified to:

$$G = \frac{(1-\upsilon^2)}{E}\pi a\sigma^2 \tag{7.2.16}$$

In Eq. (7.2.16) we can see that because Poisson ratio effect in plane strain case, the energy release rate G decreases. If we want to rewrite Eq. (7.2.14) and Eq. (7.2.15) for plane strain case then we can easily write down:

$$K_I = \sqrt{H\,G} \qquad K_{IC} = \sqrt{H\,G_C} \tag{7.2.17}$$

In Eq. (7.2.17) the modulus of elasticity E, is replaced with $H = \dfrac{E}{1-\upsilon^2}$, and we can say that Eq. (17.2.7) is valid for both plane stress and plane cases such that:

$$H = \begin{cases} E & plane\,stress \\ \dfrac{E}{1-\upsilon^2} & plane\,strain \end{cases}$$

Historically, Griffith developed what we know in this section up to Eq. (7.2.12), but later Irwin defined SIF by Eq. (7.2.13), that is, $K_I = \sigma\sqrt{\pi\,a}$ in which index I in K_I refers to mode I, in which the crack faces are opening from each other, also called (opening mode). Thereafter Irwin defined K_{II} for mode II and also K_{III} for mode III. We will study these modes in the next chapter.

Then he declared by great confidence that what is important in fracture mechanics is not stress alone, but K_I and if it approaches to fracture toughness K_{IC}, that is, ($K_I \rightarrow K_{IC}$), the fast fracture occurs. We can take K_I under control in two ways, first decreasing the applied load, that is, σ, or not using the parts with large crack in it (decreasing a) and only by these tricks we can avoid fracture. Therefore, the important parameters in fracture mechanics are K_I, K_{II} and K_{III} also combination of those. In the next section we describe one of ways that SIF can be determined.

7.3 STRESS DISTRIBUTION AND MODIFIED WESTERGAARD FUNCTIONS

As we discussed in previous section, the K_I, named as stress intensity factor (SIF) is a key parameter in fracture mechanics. One of ways that we can determine the SIF is knowledge about stress distribution around the crack tip. This fact was known to the scientists before Irwin, but finding the stress distribution around the crack tip was a formidable task.

In the Section 6.3, we derived formulas Eqs. (6.3.14), (6.3.15) and (6.3.16) for the stresses known as Koslov formulas, and he developed those in 1908 in his doctoral dissertation. However, determination of the function $\phi(z)$, and also $\psi(z)$, was and is a formidable task as we demonstrated this fact in chapter six.

Later another Russian mathematician named Muskhelishvili, developed a relationship between $\phi(\zeta)$, and $\psi(\zeta)$, by using boundary integral method, which we have proved it in equation 6.5.28. His book translated to English in 1950, and the author seen, many stress distributions around the flaws there, (not for cracks). Before him and Irwin, scientists were aware that the stress distribution around the crack tips are important, and before world war II, Westergaard wrote a paper in 1939, describing some stress functions appropriate for finding stresses around the cracks, without any proof. He named those functions $Z_I(z)$, which is still have the same notation.

After the World War II, the research on fracture mechanics started again and scientists used the Westergaard formulas, although they were given for a combined remote stress state where both $\sigma_x = \sigma$ and $\sigma_y = \sigma$ which obviously does not cover mode I fracture. Irwin was the first who suggested a change in Westergaard functions, by seeing experimental photo-elastic patterns of the stresses around the crack tips. Later his associate Sanford examined Westergaard formulas in shadow of Koslov formulas and introduced a 2nd function $Y_I(z)$, showed that Irwin's suggestion was correct.

In this section we will modify the Westergaard functions with an easier approach via closer examination of the Koslov formulas. We start with an fare assumption that in mode I fracture, along the x axis or $y = 0$, due symmetry (origin is in middle of crack) (Fig. 7.5) we can write:

$$y = 0 \Rightarrow \tau_{xy} = 0 \tag{7.3.1}$$

Now by using 6.3.16 we rewrite Eq. (7.3.1) again, that is,

$$\tau_{xy}\Big|_{y=0} = \Im\left(\bar{z}\phi''(z) + \psi''(z)\right)\Big|_{y=0} = 0 \tag{7.3.2}$$

Or simply saying:

$$\Im\left(\bar{z}\phi''(z)+\psi''(z)\right)\Big|_{y=0}=0 \tag{7.3.3}$$

This Eq. (7.3.3) necessitates that $\bar{z}\phi''(z)+\psi''(z)$ on the x axis or $y=0$, should be a real number, that is,

$$\left(\bar{z}\phi''(z)+\psi''(z)\right)\Big|_{y=0}=A \tag{7.3.4}$$

Now the important question is: what structure of $\psi(z)$ can satisfy Eq. (7.3.4)? we propose the following form:

$$\psi'(z)=\phi(z)-z\phi'(z)+Az+B \tag{7.3.5}$$

Exercise 7.3.1: Confirm that expression (7.3.5) can satisfy Eq. (7.3.4).

Solution: In Eq. (7.3.5), A is the same real number in Eq. (7.3.4) and B is a new constant. If we differentiate from Eq. (7.3.5):

$\psi''(z)=\overbrace{\phi'(z)-\phi'(z)}^{0}-z\phi''(z)+A$ which results:

$$\psi''(z)=A-z\phi''(z) \tag{7.3.6}$$

Since by assuming Eqs. (7.3.5) and (7.3.6) is always valid, it should also be valid at along the x axis or $y=0$, therefore we have:

$$\psi''(z)\Big|_{y=0}=\left(A-z\phi''(z)\right)\Big|_{y=0} \tag{7.3.7}$$

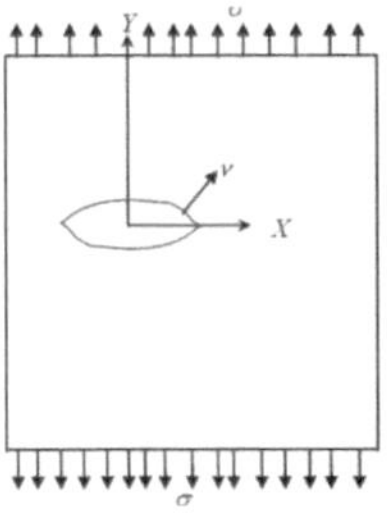

FIGURE 7.5 Symmetry.

In order to check if Eq. (7.3.4) is satisfied we only need to substitute Eq. (7.3.7) into Eq. (7.3.4) and confirm successfulness of the satisfaction.

$$\left(\bar{z}\phi''(z)+A-z\phi''(z)\right)\big|_{y=0}=A\Rightarrow\left((\bar{z}-z)\phi''(z)+A\right)\big|_{y=0}=A$$

Since, $(\bar{z}-z)=-2iy$, then we have $\left(-2iy\phi''(z)+A\right)\big|_{y=0}=A$ which provides $A=A$.

Now, we define the modified Westergaard function as follows:

$$Z_I(z)=2\phi'(z) \tag{7.3.8}$$

If we differentiate and integrate from (7.3.8) then we have:

$$Z_I'(z)=2\phi''(z) \tag{7.3.9}$$

$$\tilde{Z}_I(z)=\int Z_I(z)dz=2\phi(z) \tag{7.3.10}$$

Now we rewrite equation 6.3.14 again, that is,

$$\sigma_y=2\Re\left(\phi'(z)\right)+\Re\left(\bar{z}\phi''(z)+\psi''(z)\right) \tag{7.3.11}$$

If we substitute Eqs. (7.3.6) and (7.3.8) into Eq. (7.3.11), then we have: $\sigma_y=\Re\left(Z_I(z)\right)+\Re\left(\bar{z}\phi''(z)+A-z\phi''(z)\right)$ which results: $\sigma_y=\Re\left(Z_I(z)\right)+\Re\left((\bar{z}-z)\phi''(z)\right)+A$, since $(\bar{z}-z)=-2iy$, therefore:

$\sigma_y=\Re\left(Z_I(z)\right)-y\Re\left(2i\phi''(z)\right)+A$, and if we substitute Eq. (7.3.9) we have:

$\sigma_y=\Re\left(Z_I(z)\right)-y\Re\left(iZ_I'(z)\right)+A$, but we have $\Re\left(iZ_I'(z)\right)=-\Im\left(Z_I'(z)\right)$ (valid for every function), and then we can have:

$$\sigma_y=\Re\left(Z_I(z)\right)+y\Im\left(Z_I'(z)\right)+A \tag{7.3.12}$$

Exercise 7.3.2: Find an expression for σ_x similar to Eq. (7.3.12).

Solution: rewrite Eq. (6.3.15) again, that is,

$$\sigma_x = 2\Re\left(\phi'(z)\right) - \Re\left(\bar{z}\phi''(z) + \psi''(z)\right) \tag{7.3.13}$$

If we substitute Eqs. (7.3.6) and (7.3.8) into Eq. (7.3.11), then results: $\sigma_x = \Re\left(Z_I(z)\right) - \Re\left((\bar{z} - z)\phi''(z)\right) - A$, since $(\bar{z} - z) = -2iy$, therefore: $\sigma_x = \Re\left(Z_I(z)\right) + y\Re\left(2i\phi''(z)\right) - A$, and if we substitute Eq. (7.3.9) we have: $\sigma_y = \Re\left(Z_I(z)\right) + y\Re\left(iZ_I'(z)\right) + A$, but we have $\Re\left(iZ_I'(z)\right) = -\Im\left(Z_I'(z)\right)$ (valid for every function), and then we can have:

$$\sigma_x = \Re\left(Z_I(z)\right) - y\Im\left(Z_I'(z)\right) - A \tag{7.3.14}$$

For shear stress we need to rewrite Eq. (6.3.16) again, that is,

$$\tau_{xy} = \Im\left(\bar{z}\phi''(z) + \psi''(z)\right) \tag{7.3.15}$$

By substituting Eq. (7.3.6) into Eq. (7.3.15) we have, $\tau_{xy} = \Im\left(\bar{z}\phi''(z) + A - z\phi''(z)\right)$, since A is a real number, we have $\Im(A) = 0$, and Eq. (7.3.15) will be: $\tau_{xy} = \Im\left((\bar{z} - z)\phi''(z)\right)$ and since $(\bar{z} - z) = -2iy$, then Eq. (7.3.15) will be:

$\tau_{xy} = -y\,\Im\left(2i\,\phi''(z)\right) = -y\,\Im\left(i\,Z_I'(z)\right)$, also we have, $\Im\left(i\,Z_I'(z)\right) = \Re\left(Z_I'(z)\right)$, the Eq. (7.3.15) is:

$$\tau_{xy} = -y\,\Re\left(Z_I'(z)\right) \tag{7.3.16}$$

Equations (7.3.12), (7.3.14) and (7.3.16) give the stresses versus the modified Wetergaard stress functions. The 1st question is, what is A? The 2nd question is how we find $Z_I(z)$? When Westergaard published his paper, there was not any answer to these questions and he suggested some $Z_I(z)$ without any A in them and no proof provided. It is obvious that we have answer to both questions because in Eq. (7.1.6), we found out and expression for $\phi(z)$ which is:

$$\phi(z)=\frac{\sigma}{4}\left(2\sqrt{z^2-a^2}-z\right) \tag{7.1.6}$$

We can find $Z_I(z)$ if we substitute Eq. (7.1.6) into Eq. (7.3.8) and then we have:

$$Z_I(z)=2\phi'(z)=\frac{\sigma}{2}\left(\frac{2z}{\sqrt{z^2-a^2}}-1\right) \tag{7.3.17}$$

Differentiation of Eq. (7.3.17) gives $Z_I'(z)$, which is: $Z_I(z)=\frac{\sigma}{2}\left(\frac{2}{\sqrt{z^2-a^2}}-\frac{2z^2}{\left(\sqrt{z^2-a^2}\right)^3}\right)$, and can be simplified into:

$$Z_I'(z)=\sigma\left(\frac{1}{\sqrt{z^2-a^2}}-\frac{z^2}{\left(\sqrt{z^2-a^2}\right)^3}\right) \tag{7.3.18}$$

By substituting Eqs. (7.3.17) and (7.3.18) into Eq. (7.3.12) we have:

$$\sigma_y=\sigma\,\Re\left(\frac{z}{\sqrt{z^2-a^2}}\right)-\Re\left(\frac{\sigma}{2}\right)+y\sigma\,\Im\left(\frac{1}{\sqrt{z^2-a^2}}-\frac{z^2}{\left(\sqrt{z^2-a^2}\right)^3}\right)+A \tag{7.3.19}$$

By substituting Eqs. (7.3.17) and (7.3.18) into Eq. (7.3.14) we have:

$$\sigma_x=\sigma\,\Re\left(\frac{z}{\sqrt{z^2-a^2}}\right)-\Re\left(\frac{\sigma}{2}\right)-y\sigma\,\Im\left(\frac{1}{\sqrt{z^2-a^2}}-\frac{z^2}{\left(\sqrt{z^2-a^2}\right)^3}\right)-A \tag{7.3.20}$$

By substituting Eqs. (7.3.17) and (7.3.18) into Eq. (7.3.16) we have:

$$\tau_{xy} = -y\sigma\,\Re\left(\frac{1}{\sqrt{z^2-a^2}} - \frac{z^2}{\left(\sqrt{z^2-a^2}\right)^3}\right) \tag{21}$$

Equations Eqs. (7.3.19), (7.3.20) and (7.3.21) can answer the 2nd question, but we still do not know what is A? To answer this take the limits of the following functions when $z \to \infty$, then:

$$\lim_{z\to\infty}\frac{1}{\sqrt{z^2-a^2}} = 0,\quad \lim_{z\to\infty}\frac{z^2}{\left(\sqrt{z^2-a^2}\right)^3} = 0\quad \lim_{z\to\infty}\frac{z}{\sqrt{z^2-a^2}} = 1 \tag{7.3.22}$$

For finding A we declare that:

$$\sigma_x\big|_{z\to\infty} = 0 \tag{7.3.23}$$

If we implement Eq. (7.3.23) in Eqs. (17.3.9), (7.3.20) and (7.3.21) with considering the limits in Eq. (7.3.22) we can write:

$$\sigma_x\big|_{z\to\infty} = \sigma\,\Re\left(\lim_{z\to\infty}\frac{z}{\sqrt{z^2-a^2}}\right) - \Re\left(\frac{\sigma}{2}\right)$$
$$-y\sigma\,\Im\left(\lim_{z\to\infty}\frac{1}{\sqrt{z^2-a^2}} - \lim_{z\to\infty}\frac{z^2}{\left(\sqrt{z^2-a^2}\right)^3}\right) - A$$

and simplifies to:

$$0 = \sigma - \frac{\sigma}{2} - A \text{ or: } A = \frac{\sigma}{2} \tag{7.3.24}$$

Substituting Eq. (7.3.24) into Eqs. (7.3.19), (7.3.20) and (7.3.21) provides the stress formulas that contains full information (both the questions answered).

$$\sigma_y = \left[\Re\left(\frac{z}{\sqrt{z^2 - a^2}} \right) + y\,\Im\left(\frac{1}{\sqrt{z^2 - a^2}} - \frac{z^2}{\left(\sqrt{z^2 - a^2}\right)^3} \right) \right] \sigma \quad (7.3.25)$$

$$\sigma_x = \left[\Re\left(\frac{z}{\sqrt{z^2 - a^2}} \right) - y\,\Im\left(\frac{1}{\sqrt{z^2 - a^2}} - \frac{z^2}{\left(\sqrt{z^2 - a^2}\right)^3} - 1 \right) \right] \sigma \quad (7.3.26)$$

And for the shear stress the Eq. (7.3.21) is valid. Equations (7.3.21), (7.3.25) and (7.3.26) gives the stress formula. Historically the term 1 in Eq. (7.3.26) cleverly suggested by Irwin without proof, when he used the Westergaard formulas. In the next chapter we use these formulas to study stress distribution near the crack tips in various modes.

KEYWORDS

- **Griffith theory**
- **Koslov formulas**
- **linear elastic fracture mechanic**
- **stress intensity factor**
- **Westergaard functions**

CHAPTER 8

DESCRIBING THREE MODES OF FRACTURE

CONTENTS

8.1 CRACK TIP STRESSES IN MODE I FRACTURE

In previous chapter, it was seen that how difficult the calculation of the stresses σ_y, σ_x and τ_{xy} can be. Those stress formulas were given by Eqs. (7.3.21), (7.3.25) and (7.326). Fortunately, we are interested in stress distribution around the crack tips, since it is only this type of distribution that can lead us to K_I or K_{IC}. Therefore, in Eqs. (7.3.21), (7.3.25) and (7.3.26), we transfer the coordinate of the origin to the crack tip (Fig. 8.1), in mathematical term new variable $z_0 = z - a = r\,e^{i\theta}$ will be entered into the discussions, rewriting this provides:

$$z_0 = z - a = r\,e^{i\theta} \Rightarrow z = a + r\,e^{i\theta} \tag{8.1.1}$$

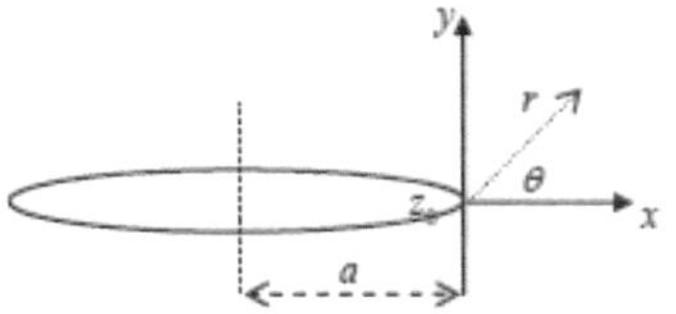

FIGURE 8.1 Crack tip.

Before we start calculations, we should emphasize why the stresses around crack tip and particularly, the Westergaard functions are that important. We can answer to this question by rewriting Eq. (7.3.17) again which is:

$$Z_I(z)=\frac{\sigma}{2}\left(\frac{2z}{\sqrt{z^2-a^2}}-1\right) \tag{7.3.17}$$

If we multiply both sides of the above equation into $\sqrt{z-a}$, then we have:

$\sqrt{z-a}\,Z_I(z)=\frac{\sigma}{2}\left(\frac{2z\sqrt{z-a}}{\sqrt{z^2-a^2}}-\sqrt{z-a}\right)$ and then it can be simplified to:

$\sqrt{z-a}\,Z_I(z)=\frac{\sigma}{2}\left(\frac{2z}{\sqrt{z+a}}-\sqrt{z-a}\right)$, then we will take a limit as $z\rightarrow a$ and we have:

$$\lim_{z\rightarrow a}\sqrt{z-a}\,Z_I(z)=\frac{\sigma}{2}\left(\lim_{z\rightarrow a}\frac{2z}{\sqrt{z+a}}-\lim_{z\rightarrow a}\sqrt{z-a}\right) \tag{8.1.2}$$

which yields to:

$$\lim_{z\rightarrow a}\sqrt{z-a}\,Z_I(z)=\frac{\sigma}{2}\left(\frac{2a}{\sqrt{2a}}\right)=\sigma\sqrt{\frac{a}{2}} \tag{8.1.3}$$

We saw in Eq. (7.2.13), that SIF in a very simple case is:

$$K_I=\sigma\sqrt{\pi a} \tag{8.1.4}$$

Now we multiply both sides of Eq. (8.1.3) into $\sqrt{2\pi}$ and we have:

$$\sqrt{2\pi}\lim_{z\rightarrow a}\sqrt{z-a}\,Z_I(z)=\sigma\sqrt{\pi a} \tag{8.1.5}$$

By comparing Eqs. (8.1.4) and (8.1.5) we can easily find that:

$$K_I=\sqrt{2\pi}\lim_{z\rightarrow a}\sqrt{z-a}\,Z_I(z) \tag{8.1.6}$$

It is Eq. (8.1.6) that says, why the stress distribution around the crack tip and analytical methods for its determination, are so important. Although we have derived Eq. (8.1.6) for a very simple case in mode I, scientists believe that it can be modified for any cases by a correction factor. This is motivated mathematicians to find the stress distributions near the crack tip and ignore the higher order terms, so that finally it leads them to SIF.

First we need to interpret what is vicinity of the crack tip? We need to examine Eq. (8.1.1) and vicinity of crack tip $z \cong a$, since in Eq. (8.1.1), only the region $r << a$ identified as crack tip neighborhood, we can approximate:

$$z + a \cong 2a \tag{8.1.7}$$

Therefore in Eqs. (7.3.25) and (7.3.26), the terms like $\Re\left(\frac{z}{\sqrt{z^2 - a^2}}\right)$, can be approximated like this:

$\Re\left(\frac{z}{\sqrt{z^2 - a^2}}\right) = \Re\left(\frac{a + re^{i\theta}}{\sqrt{z-a}\sqrt{z+a}}\right) \cong \Re\left(\frac{a}{\sqrt{re^{i\theta}}\sqrt{2a}}\right)$, the reason is that for $r << a$, we have $a + re^{i\theta} \cong a$, further simplification gives:

$\Re\left(\frac{z}{\sqrt{z^2 - a^2}}\right) \cong \Re\left(\sqrt{\frac{a}{2r}}e^{-i\theta/2}\right) \cong \Re\left(\sqrt{\frac{a}{2r}}\left(\cos\frac{\theta}{2} - i\sin\frac{\theta}{2}\right)\right)$, finally we can write:

$$\Re\left(\frac{z}{\sqrt{z^2 - a^2}}\right) \cong \sqrt{\frac{a}{2r}}\cos\frac{\theta}{2} \tag{8.1.8}$$

It is obvious that several expressions like (8.1.8) should be derived, such that, the Eqs. (7.3.21), (7.3.25) and (7.3.26) can be converted into the forms that are appropriate for crack tip region. The next approximation is:

$y\Im\left(\frac{1}{\sqrt{z^2 - a^2}}\right) = y\Im\left(\frac{1}{\sqrt{z-a}\sqrt{z+a}}\right) \cong y\Im\left(\frac{1}{\sqrt{re^{i\theta}}\sqrt{2a}}\right)$ and since $y = r\sin\theta$, we can write: $y\Im\left(\frac{1}{\sqrt{z^2 - a^2}}\right) \cong \frac{r\sin\theta}{\sqrt{2ar}}\Im\left(\cos\frac{\theta}{2} - i\sin\frac{\theta}{2}\right) \cong -\sqrt{\frac{r}{2a}}\sin\theta\sin\frac{\theta}{2}$ and since around the crack tip $r << a$, we can say $\sqrt{\frac{r}{2a}} \cong 0$, therefore we can write:

$$y\,\Im\left(\frac{1}{\sqrt{z^2-a^2}}\right)\cong 0 \tag{8.1.9}$$

$$y\,\Im\left(\frac{z^2}{\left(\sqrt{z^2-a^2}\right)^3}\right)=y\,\Im\left(\frac{z^2}{\left(\sqrt{z-a}\right)^3\left(\sqrt{z+a}\right)^3}\right)$$

$$\cong y\,\Im\left(\frac{a^2}{\left(\sqrt{r\,e^{i\theta}}\right)^3\left(\sqrt{2a}\right)^3}\right)$$

Exercise 8.1.1: Since $y=r\sin\theta$, simplify the above expressions.

Solution:

$$y\,\Im\left(\frac{z^2}{\left(\sqrt{z^2-a^2}\right)^3}\right)\cong\Im\left(\frac{r\,a^2\sin\theta}{\left(\sqrt{r\,e^{i\theta}}\right)^3\left(\sqrt{2a}\right)^3}\right)\cong\Im\left(\frac{r\,a^2\sin\theta}{2\sqrt{2}\,r\sqrt{r}\,a\sqrt{a}\,e^{3i\theta/2}}\right)$$

also from trigonometry we have, $\sin\theta=2\sin\frac{\theta}{2}\cos\frac{\theta}{2}$ then above expression becomes:

$$y\,\Im\left(\frac{z^2}{\left(\sqrt{z^2-a^2}\right)^3}\right)\cong\sqrt{\frac{a}{2r}}\sin\frac{\theta}{2}\cos\frac{\theta}{2}\,\Im\left(e^{-3i\theta/2}\right)\cong$$

$$\sqrt{\frac{a}{2r}}\sin\frac{\theta}{2}\cos\frac{\theta}{2}\,\Im\left(\cos\frac{3\theta}{2}-i\sin\frac{3\theta}{2}\right)$$

$$y\,\Re\left(\frac{z^2}{\left(\sqrt{z^2-a^2}\right)^3}\right)\cong\sqrt{\frac{a}{2r}}\sin\frac{\theta}{2}\cos\frac{\theta}{2}\,\Re\left(e^{-3i\theta/2}\right)\cong$$

$$\sqrt{\frac{a}{2r}}\sin\frac{\theta}{2}\cos\frac{\theta}{2}\,\Re\left(\cos\frac{3\theta}{2}-i\sin\frac{3\theta}{2}\right)$$

From the above expression we can find two useful formulas which are:

$$y\,\Im\left(\frac{z^2}{\left(\sqrt{z^2-a^2}\right)^3}\right) \cong -\sqrt{\frac{a}{2r}}\sin\frac{\theta}{2}\cos\frac{\theta}{2}\sin\frac{3\theta}{2} \qquad (8.1.10a)$$

$$y\,\Re\left(\frac{z^2}{\left(\sqrt{z^2-a^2}\right)^3}\right) \cong \sqrt{\frac{a}{2r}}\sin\frac{\theta}{2}\cos\frac{\theta}{2}\cos\frac{3\theta}{2} \qquad (8.1.10b)$$

If we substitute the Eqs. (8.1.8), (8.1.9), (8.1.10a) and (8.1.10b) into Eqs. (7.3.25), (7.3.26) and (7.3.21), then we have:

$$\sigma_y = \sigma\sqrt{\frac{a}{2r}}\cos\frac{\theta}{2}\left(1-\sin\frac{\theta}{2}\sin\frac{3\theta}{2}\right) \qquad (8.1.11)$$

$$\sigma_x = \sigma\sqrt{\frac{a}{2r}}\cos\frac{\theta}{2}\left(1+\sin\frac{\theta}{2}\sin\frac{3\theta}{2}\right)-\sigma \qquad (8.1.12)$$

$$\tau_{xy} = \sigma\sqrt{\frac{a}{2r}}\sin\frac{\theta}{2}\cos\frac{\theta}{2}\cos\frac{3\theta}{2} \qquad (8.1.13)$$

Equations (8.1.11), (8.1.12) and (8.1.13) gives the stress distribution in mode I, in vicinity of the crack tip. It is obvious that the higher order terms are ignored but this ignorance around the crack tip vicinity is justifiable. Since K_I and $Z_I(z)$ are related via (8.1.6) it is obvious that K_I and stresses are also related. The stresses can then be computed by FEM and K_I also related to it. To convert Eqs. (8.1.11) and (8.1.12) in terms of K_I we need:

$$\sigma\sqrt{\frac{a}{2r}} = \sigma\sqrt{\frac{\pi a}{2\pi r}} = \frac{K_I}{\sqrt{2\pi r}} \qquad (8.1.14)$$

Substituting Eq. (8.1.14) into Eqs. (8.1.11), (8.1.12) and (8.1.13) provides new set of formulas:

$$\sigma_y = \frac{K_I}{\sqrt{2\pi r}} \cos\frac{\theta}{2}\left(1 - \sin\frac{\theta}{2}\,\sin\frac{3\theta}{2}\right) \tag{8.1.15}$$

$$\sigma_x = \frac{K_I}{\sqrt{2\pi r}} \cos\frac{\theta}{2}\left(1 + \sin\frac{\theta}{2}\,\sin\frac{3\theta}{2}\right) - \sigma \tag{8.1.16}$$

$$\tau_{xy} = \frac{K_I}{\sqrt{2\pi r}} \sin\frac{\theta}{2}\cos\frac{\theta}{2}\cos\frac{3\theta}{2} \tag{8.1.17}$$

Formulas (8.1.15), (8.1.16) and (8.1.17) are found based on the modified Westergaard functions. In many books term $-\sigma$ does not exist, the reason is that they have used the original Westergaard stress functions published in 1939.

Later when scientists who studied the stress patterns around the crack, found that existing stress distribution formulas on that time, does not match with the experimental observations from photo elasticity. Irwin explained this controversy by saying that the original Westergaard function is given for a case $\sigma_y = \sigma_x = \sigma$, but mode I fracture is for the case when $\sigma_y = \sigma$ and $\sigma_x = 0$. Therefore, any stress formulas on that time or even in some books now, cannot be correct. He cleverly suggested that term $-\sigma$ should be added to Eq. (8.1.16) so the condition $\sigma_x = 0$ can be satisfied. Then scientist investigated and found out that using Eq. (8.1.16) as suggested Irwin completely matches with the experimental results obtained from the photo-elasticity. This fact is now proved theoretically in this section.

Now if we multiply both sides of Eq. (8.1.15) into $\sqrt{2\pi r}$ and take the limit when $r \to 0$ and also $\theta \to 0$ then we can find a very interesting formula like this:

$$K_I = \lim_{r,\theta \to 0} \sqrt{2\pi r}\,\sigma_y \tag{8.1.18}$$

Equation (8.1.16) provides an important fact in fracture mechanics, by saying that in mode I, the stress distribution around the tip is closely related to SIF. This is why the stress distribution in vicinity of the crack tips is as important as determination of SIF. Not only analytical method but also numerical methods like FEM first calculate σ_y around the crack and then determine K_I or SIF.

Another interesting point is in Eqs. (8.1.11), (8.1.12) and (8.1.13), if we want to find the stresses at $r \to \infty$ that is very far from the crack vicinity we have:

$$\sigma_y\Big|_{r\to\infty} = 0 \qquad \sigma_x\Big|_{r\to\infty} = -\sigma \qquad \tau_{xy}\Big|_{r\to\infty} = 0 \tag{8.1.19}$$

It is obvious that $\sigma_y\Big|_{r\to\infty} = \sigma$ and $\sigma_x\Big|_{r\to\infty} = 0$, this should not surprise us that Eq. (8.1.19) is not correct because all the stress formulas in this section are valid for crack vicinity only.

8.2 WESTERGAARD FUNCTIONS AND MODE II FRACTURE

Before we enter to mode II fracture (Fig. 8.2), we should say that our purpose is studying the stress distribution around the crack. The formulas in this section also are not valid $r >> a$ or also $r \to \infty$. They will be valid only in crack tip vicinity or $r \cong a$ and the formulas will be valid for determination of SIF in mode II or K_{II}. This mode is also known as slipping mode and is caused by remote shear stress as Irwin mentioned the remote stress $\tau_{xy}\Big|_{z\to\infty} = \tau$ causes the crack to slide on its own plane if the shear stress is too much the crack start to grow and fast fracture occurs. All we know about the remote shear stress is from Eq. (8.1.8), and we write them here again.

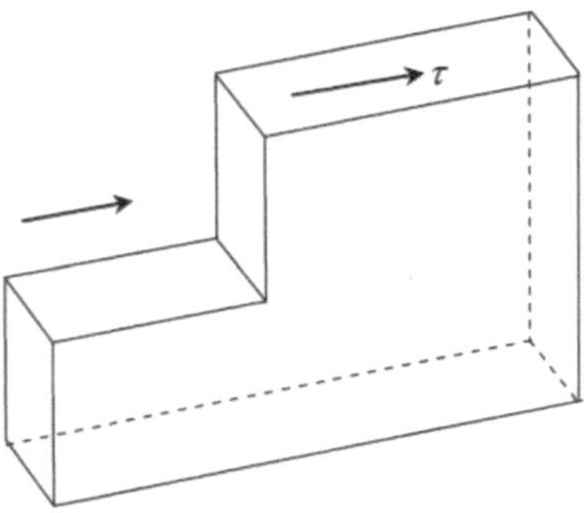

FIGURE 8.2 Mode II fracture.

In this section we are interested in $\psi'(z)\big|_{z\to a}$ and also $\phi(z)\big|_{z\to a}$ only, but Eq. (8.2.1) helps that we find these function easier. Then we show that the Westergaard formula:

$$\psi'(z)\big|_{z\to\infty} = i\,\tau\,z \qquad \phi(z)\big|_{z\to\infty} = 0 \tag{8.2.1}$$

which can be justified. If we want to know about $\phi(\zeta)$ for mode II fracture, we need to look at Eq. (6.6.2), and there if we set $\sigma = 0$ we find:

$$\phi_{II}(\zeta) = \sum_{n=1}^{\infty} A_n\,\zeta^{1-2n} \tag{8.2.2}$$

It is obvious that the above function for $\phi_{II}(\zeta)$, can be accepted because $\phi_{II}(\zeta)\big|_{\zeta\to\infty} = 0$. For the reasons that we was discussed in Eq. (8.1.6), we need to keep only one term in Eq. (8.2.2) which is for $n = 1$, that is,

$$\phi_{II}(\zeta) = \frac{|A_1|}{\zeta} \tag{8.2.3}$$

In the above equation absolute value sign means for positive shear stresses. The question is how can we find A_1? The answer again is given in article 6.6 by Eq. ((6.6.12)), if we set $\sigma = 0$, we have a formula for mode II.

$$\psi_{II}(\zeta)\big|_{\zeta\to\infty} = -A_1\,\zeta \tag{8.2.4}$$

Moreover, for a crack with the length $2a$, the mapping function is given by Eq. (8.2.3) which is:

$$z = \frac{a}{2}\left(\zeta + \frac{1}{\zeta}\right) \tag{8.2.5}$$

If we substitute Eq. (8.2.5) into Eq. (8.2.1) we can write:

$$\psi'(z)\big|_{z\to\infty} = i\,\tau\frac{a}{2}\left(\zeta + \frac{1}{\zeta}\right) \tag{8.2.6}$$

Now in Eq. (8.2.6) we set $\zeta \to \infty$ and according to the notation in Section 6.6, we replace $\psi'(z)\big|_{z\to\infty}$ with the $\psi_{II}(\zeta)\big|_{\zeta\to\infty}$, then we have:

$$\psi_{II}(\zeta)\big|_{\zeta\to\infty} = i\,\tau\frac{a}{2}\zeta \tag{8.2.7}$$

If we compare Eq. (8.2. 7) with Eq. (8.2.4) then we have:

$$A_1 = -i\,\tau\frac{a}{2} \qquad |A_1| = i\,\tau\frac{a}{2} \tag{8.2.8}$$

If we substitute Eq. (8.2.8) into Eq. (8.2.3), then we have:

$$\phi_{II}(\zeta) = i\,\tau\frac{a}{2\zeta} \tag{8.2.9}$$

From Eq. (7.1.5), we saw that $\frac{1}{\zeta} = \frac{z-\sqrt{z^2-a^2}}{a}$ and then (9) will change in terms of z i.e.

$$\phi_{II}(z) = i\frac{\tau}{2}\left(z-\sqrt{z^2-a^2}\right) \tag{8.2.10}$$

Equation (8.2.10) is the basis, for mode II fracture and in this section we show that from (8.2.10) we can find the modified Westergaard function. To do this we refer to Koslov formulas and write Eq. (6.3.14) herein again.

$$\sigma_y = \Re\left(2\phi'(z)+\bar{z}\phi''(z)+\psi''(z)\right) \tag{8.2.11}$$

This section for the first time finds the modified Westergaard function for mode II fracture and this based on the assumption that over the x axis there is not normal stress.

$$\sigma_y\big|_{y=0} = 0 \tag{8.2.12}$$

According to Eqs. (8.2.11) and (8.2.12) we can write that:

$$2\phi'(z)+\bar{z}\phi''(z)+\psi''(z)\Big|_{y=0}=iB \tag{8.2.13}$$

Similar to the reasoning put forward in Section 7.2, in order for Eq. (8.2.13) to be satisfied, the function $\psi'(z)$ should have the following form: (B is real and iB is imaginary).

$$\psi'(z)=-\phi(z)-z\phi'(z)+iBz \tag{8.2.14}$$

Exercise 8.2.1: Confirm that expression (8.2.14) can satisfy Eq. 8.2.13).

Solution: If we differentiate from Eq. (8.2.14):

$\psi''(z)=\overbrace{-\phi'(z)-\phi'(z)}^{-2\phi'(z)}-z\phi''(z)+iB$ which results:

$$\psi''(z)=iB-z\phi''(z)-2\phi'(z) \tag{8.2.15}$$

Since by assuming Eqs. (8.2.14) and (8.2.15) is always valid, it should also be valid along the x axis or $y=0$, therefore we have:

$$\psi''(z)\Big|_{y=0}=\left(iB-z\phi''(z)-2\phi'(z)\right)\Big|_{y=0}$$

In order to check if Eq. (8.2.13) is satisfied we only need to substitute the above into Eq. (8.2.13) and confirm successfulness of the satisfaction.

$$\left(\bar{z}\phi''(z)+iB-z\phi''(z)\overbrace{-2\phi'(z)+2\phi'(z)}^{0}\right)\Big|_{y=0}$$
$$=iB\Rightarrow\left((\bar{z}-z)\phi''(z)+iB\right)\Big|_{y=0}=iB$$

Since, $(\bar{z}-z)=-2iy$, then we have $\left(-2iy\phi''(z)+iB\right)\Big|_{y=0}=iB$ which provides $iB=iB$.

Now, we define the modified Westergaard function as follows:

$$Z_{II}(z)=2\phi'(z) \tag{8.2.16a}$$

If we differentiate from (8.2.16a) then we have:

$$Z_{II}'(z) = 2\phi''(z) \tag{8.2.16b}$$

In general form as well, $\sigma_y = \Re\left(\bar{z}\phi''(z) + iB - z\phi''(z) \overbrace{+2\phi'(z) - 2\phi'(z)}^{0}\right)$

which results: $\sigma_y = \underbrace{\Re(iB)}_{0} + \Re((\bar{z} - z)\phi''(z))$, since $(\bar{z} - z) = -2iy$, therefore:

$$\sigma_y = \Re(-2iy\phi''(z)) \tag{8.2.17}$$

Substituting Eq. (8.2.16b) into Eq. (8.2.17) gives $\sigma_y = \Re(-iyZ_{II}'(z))$ and since $\Re(-iZ_{II}'(z)) = \Im(Z_{II}'(z))$

Then we can write:

$$\sigma_y = y\Im(Z_{II}'(z)) \tag{8.2.18}$$

Exercise 8.2.2: Find an expression for σ_x similar to Eq. (8.2.18).

Solution: rewrite Eq. (6.3.15) again, that is,

$$\sigma_x = \Re(2\phi'(z) - \bar{z}\phi''(z) - \psi''(z)) \tag{8.2.19}$$

If we substitute Eq. (8.2.15) into Eq. (8.2.19), then results: $\sigma_x = \Re(2\phi'(z) - \bar{z}\phi''(z) - iB + z\phi''(z) + 2\phi'(z))$, since $(\bar{z} - z) = -2iy$ and also according to Eq. (8.2.16a):

$$\sigma_x = \Re(2Z_{II}(z)) + y\Re(2i\phi''(z)) - \overbrace{\Re(iB)}^{0} \tag{8.2.20}$$

and if we substitute Eq. (8.2.16b) we have: $\sigma_x = \Re(Z_{II}(z)) + y\Re(iZ_{II}'(z))$, but we have $\Re(iZ_{II}'(z)) = -\Im(Z_{II}'(z))$ (valid for every function), and then we can have:

$$\sigma_x = 2\Re(Z_{II}(z)) - y\Im(Z_{II}'(z)) \tag{8.2.21}$$

The important part in mode II fracture is determination of the shear stresses around the crack because it may be a dominant stress. For shear stress we need to rewrite Eq. (6.3.16) again, that is,

$$\tau_{xy} = \Im\left(\bar{z}\phi''(z) + \psi''(z)\right) \tag{8.2.22}$$

By substituting Eq. (8.2.15) into Eq. (8.2.22) we have, $\tau_{xy} = \Im\left(\bar{z}\phi''(z) + iB - z\phi''(z) - 2\phi'(z)\right)$, since iB is a imaginary number, we have $\Im(iB) = B$, and then Eq. (8.2.22) will be: $\tau_{xy} = \Im\left((\bar{z} - z)\phi''(z) - 2\phi'(z)\right) + B$ also $(\bar{z} - z) = -2i\,y$, and this changes Eq. (8.2.22) into:

$$\tau_{xy} = \Im\left(-2i\,y\,\phi''(z) - 2\phi'(z)\right) + B$$

Substituting the definitions (8.2.16a) and (8.2.16b) into the above expression changes it to:

$$\tau_{xy} = \Im\left(-i\,y\,Z'_{II}(z) - Z_{II}(z)\right) + B$$

From the complex variable theory we have, $\Im\left(i\;Z'_{II}(z)\right) = \Re\left(\;Z'_{II}(z)\right)$, then the above expression will change to:

$$\tau_{xy} = -\,y\,\Re\left(\;Z'_{II}(z)\right) - \Im\left(Z_{II}(z)\right) + B \tag{8.2.23}$$

The question is what is the real constant B? In original Westergaard function the term B does not exist. Finding B requires a lengthy procedure, which will be discussed in the next section.

8.3 STRESSES AROUND CRACK TIP IN MODE II FRACTURE

First we need to rewrite 8.2.10 again, that is,

$$\phi(z) = i\,\frac{\tau}{2}\left(z - \sqrt{z^2 - a^2}\right) \tag{8.2.10}$$

We take 1st and 2nd derivative of the above equation and we have:

$$\phi'(z) = i\,\frac{\tau}{2}\left(1 - \frac{z}{\sqrt{z^2 - a^2}}\right) \tag{8.3.1}$$

$$\phi''(z) = i\frac{\tau}{2}\left(\frac{-1}{\sqrt{z^2-a^2}} + \frac{z^2}{\left(\sqrt{z^2-a^2}\right)^3}\right) \tag{8.3.2}$$

Using (8.2.16a) and (8.2.16b), for definition of Westergaard functions, from (8.3.1) and (8.3.2) we have:

$$Z_{II}(z) = \left(1 - \frac{z}{\sqrt{z^2-a^2}}\right) i\tau \tag{8.3.3}$$

$$Z'_{II}(z) = \left(\frac{-1}{\sqrt{z^2-a^2}} + \frac{z^2}{\left(\sqrt{z^2-a^2}\right)^3}\right) i\tau \tag{8.3.4}$$

In order to find the stresses from Eqs. (8.3.3) and (8.3.4) we need to evaluate following terms:

$\Im\left(Z'_{II}(z)\right) = \Im\left[\left(\frac{-1}{\sqrt{z^2-a^2}} + \frac{z^2}{\left(\sqrt{z^2-a^2}\right)^3}\right) i\tau\right]$ using the rule $\Im(if) = \Re(f)$ then we have:

$$\Im\left(Z'_{II}(z)\right) = \Re\left[\left(\frac{-1}{\sqrt{z^2-a^2}} + \frac{z^2}{\left(\sqrt{z^2-a^2}\right)^3}\right) \tau\right] \tag{8.3.5}$$

The next term that should be looked at is:

$\Re\left(Z_{II}(z)\right) = \Re\left[\left(1 - \frac{z}{\sqrt{z^2-a^2}}\right) i\tau\right]$ and if we use the rule $\Re(if) = -\Im(f)$ then we have:

$$\Re\left(Z_{II}(z)\right) = -\Im\left[\left(1 - \frac{z}{\sqrt{z^2-a^2}}\right) \tau\right] \tag{8.3.6}$$

We also need to evaluate this terms:

$\Re\left(Z_{II}^{'}(z)\right)=\Re\left[\left(\frac{-1}{\sqrt{z^{2}-a^{2}}}+\frac{z^{2}}{\left(\sqrt{z^{2}-a^{2}}\right)^{3}}\right)i\tau\right]$ using the rule $\Re(i\,f)=-\Im(f)$ then we have:

$$\Re\left(Z_{II}^{'}(z)\right)=-\Im\left[\left(\frac{-1}{\sqrt{z^{2}-a^{2}}}+\frac{z^{2}}{\left(\sqrt{z^{2}-a^{2}}\right)^{3}}\right)\tau\right] \tag{8.3.7}$$

The last expression that should be analyzed is:

$\Im\left(Z_{II}(z)\right)=\Im\left[\left(1-\frac{z}{\sqrt{z^{2}-a^{2}}}\right)i\tau\right]$ using the rule $\Im(i\,f)=\Re(f)$ then we have:

$$\Im\left(Z_{II}(z)\right)=\Re\left[\left(1-\frac{z}{\sqrt{z^{2}-a^{2}}}\right)\tau\right] \tag{8.3.8}$$

Substituting Eqs. (8.3.6) and (8.3.5) into Eq. (8.2.21) gives:

$\sigma_{x}=-2\Im\left[\left(1-\frac{z}{\sqrt{z^{2}-a^{2}}}\right)\tau\right]-y\Re\left[\left(\frac{-1}{\sqrt{z^{2}-a^{2}}}+\frac{z^{2}}{\left(\sqrt{z^{2}-a^{2}}\right)^{3}}\right)\tau\right]$ and $\Im(\tau)=0$ it yields:

$$\sigma_{x}=2\Im\left(\frac{z\tau}{\sqrt{z^{2}-a^{2}}}\right)-y\Re\left[\left(\frac{-1}{\sqrt{z^{2}-a^{2}}}+\frac{z^{2}}{\left(\sqrt{z^{2}-a^{2}}\right)^{3}}\right)\tau\right] \tag{8.3.9}$$

If $z\to\infty$ in Eq. (8.3.9) then $\sigma_{x}=0$, which proves it is correct. By substituting Eq. (8.3.5) into Eq. (8.2.18):

$$\sigma_{y}=y\Re\left[\left(\frac{-1}{\sqrt{z^{2}-a^{2}}}+\frac{z^{2}}{\left(\sqrt{z^{2}-a^{2}}\right)^{3}}\right)\tau\right] \tag{8.3.10}$$

If $z \to \infty$ in Eq. (8.3.10) then $\sigma_y = 0$, which proves it is correct. By substituting Eqs. (8.3.7) and (8.3.8) into Eq. (8.2.23) we can find the shear stresses, that is,:

$$\tau_{xy} = y\,\Im\left[\left(\frac{-1}{\sqrt{z^2-a^2}} + \frac{z^2}{\left(\sqrt{z^2-a^2}\right)^3}\right)\tau\right] - \Re\left[\left(1-\frac{z}{\sqrt{z^2-a^2}}\right)\tau\right] + B$$

In the above equation if we want to calculate B, it is required to take limit at $z \to \infty$ such that $\tau_{xy}\big|_{z\to\infty} = \tau$ and considering that $\lim_{z\to\infty} \frac{z}{\sqrt{z^2-a^2}} = 1$, we have $B = \tau$ and if we substitute it will be simplified to:

$$\tau_{xy} = y\,\Im\left[\left(\frac{-1}{\sqrt{z^2-a^2}} + \frac{z^2}{\left(\sqrt{z^2-a^2}\right)^3}\right)\tau\right] - \Re\left(\frac{z\tau}{\sqrt{z^2-a^2}}\right) \quad (8.3.11)$$

Equations (8.3.9), (8.3.10) and (8.3.11) are the general formulas for the stresses in mode II fracture. If we want to approximate them in vicinity of the crack tip or $z \to a$, their shape will change. For this purpose we need to refer to Eq. (8.3.1) and bring many equations to this section and we rewrite Eqs. (8.1.8), (8.1.9) and (8.1.10a) again, that is,

$$\Re\left(\frac{z}{\sqrt{z^2-a^2}}\right) \cong \sqrt{\frac{a}{2r}}\cos\frac{\theta}{2} \qquad r << a \quad (8.3.12)$$

$$\Im\left(\frac{1}{\sqrt{z^2-a^2}}\right) \cong 0 \; r << a \quad (8.3.13)$$

$$y\,\Im\left(\frac{z^2}{\left(\sqrt{z^2-a^2}\right)^3}\right) \cong -\sqrt{\frac{a}{2r}}\sin\frac{\theta}{2}\cos\frac{\theta}{2}\sin\frac{3\theta}{2} \qquad r << a \quad (8.3.14)$$

If we substitute Eqs. (8.3.12), (8.3.13) and (8.3.14) into Eq. (8.3.11) we have:

$$\tau_{xy} = \tau\sqrt{\frac{a}{2r}}\left(\cos\frac{\theta}{2} - \sin\frac{\theta}{2}\cos\frac{\theta}{2}\sin\frac{3\theta}{2}\right) \quad (8.3.15)$$

Rewriting Eq. (8.3.10b) again:

$$y\,\Re\left(\frac{z^2}{\left(\sqrt{z^2-a^2}\right)^3}\right)\cong\sqrt{\frac{a}{2r}}\sin\frac{\theta}{2}\cos\frac{\theta}{2}\cos\frac{3\theta}{2} \tag{8.3.16}$$

and we substitute in Eq. (8.3.10) assuming $\Re\left(\frac{1}{\sqrt{z^2-a^2}}\right)\cong 0$ we have:

$$\sigma_y=\tau\sqrt{\frac{a}{2r}}\sin\frac{\theta}{2}\cos\frac{\theta}{2}\cos\frac{3\theta}{2} \tag{8.3.17}$$

In order to find σ_x, we need to evaluate another term, which is: $\Im\left(\frac{z}{\sqrt{z^2-a^2}}\right)\cong\Im\left(\sqrt{\frac{a}{2r}}e^{-iq/2}\right)\cong\Im\left(\sqrt{\frac{a}{2r}}\left(\cos\frac{q}{2}-i\sin\frac{q}{2}\right)\right)$, finally we can write:

$$\Im\left(\frac{z}{\sqrt{z^2-a^2}}\right)\cong-\sqrt{\frac{a}{2r}}\sin\frac{\theta}{2}\quad r<<a \tag{8.3.18}$$

Substituting Eqs. (8.3.18) and (8.3.16) into Eq. (8.3.9) gives the σ_x

$$\sigma_x=-\tau\sqrt{\frac{a}{2r}}\left(2\sin\frac{\theta}{2}+\sin\frac{\theta}{2}\cos\frac{\theta}{2}\cos\frac{3\theta}{2}\right) \tag{8.3.19}$$

In mode I fracture, the stress distribution around the tip, provided the access to K_I, similarly in mode II fracture the same statement is valid and we have access to K_{II}, defined by:

$$K_{II}=\tau\sqrt{\pi a} \tag{8.3.20}$$

In Eq. (8.3.20) K_{II} is calculated for the simplest case and σ is replaced with τ. If we substitute Eq. (8.3.20) into Eqs. (8.3.15), (8.3.17) and (8.3.19) then we have:

$$\sigma_x=\frac{-K_{II}}{\sqrt{2\pi r}}\sin\frac{\theta}{2}\left(2+\cos\frac{\theta}{2}\cos\frac{3\theta}{2}\right) \tag{8.3.21}$$

$$\sigma_y = \frac{K_{II}}{\sqrt{2\pi r}} \sin\frac{\theta}{2} \cos\frac{\theta}{2} \cos\frac{3\theta}{2} \tag{8.3.22}$$

$$\tau_{xy} = \frac{K_{II}}{\sqrt{2\pi r}} \cos\frac{\theta}{2}\left(1 - \sin\frac{\theta}{2} \sin\frac{3\theta}{2}\right) \tag{8.3.23}$$

We multiply both sides of Eq. (8.3.23) into $\sqrt{2\pi r}$ and also set $\theta = 0$ then we have:

$$K_{II} = \lim_{r\to 0}\left(\sqrt{2\pi r}\ \tau_{xy}\big|_{\theta=0}\right) \tag{8.3.24}$$

We can conclude that, SIF in mode II fracture can be obtained from Eq. (8.3.24). Therefore, shear stress in vicinity of crack should be studied and this can be done by FEM or photo-elasticity.

8.4 MODE III FRACTURE, ANTI-PLANE SHEAR

There may be loading normal to the plane which causes anti-plane shear. The remote stresses produced in modes I, II and III are shown in Fig. 8.3, it can be seen that the shear stress in mode III is normal to the plane and the z dimension is also involved. In this mode as can be seen in Fig. 8.4 it looks like the body is cut by a scissor. The equilibrium equation in absence of σ_z will be:

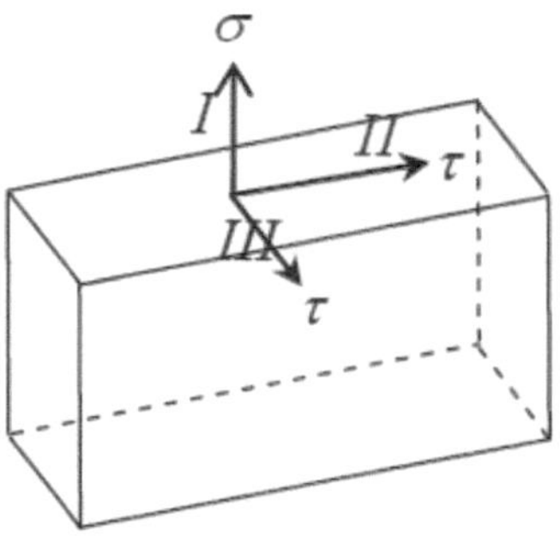

FIGURE 8.3 Shear stress in mode III.

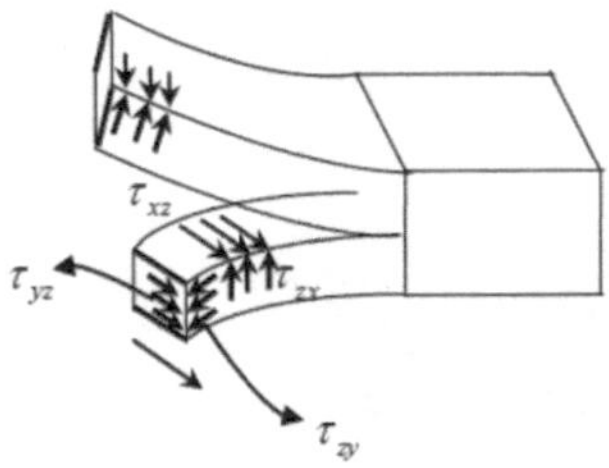

FIGURE 8.4 Anti-plane shear.

$$\frac{\partial \tau_{xz}}{\partial x}+\frac{\partial \tau_{yz}}{\partial y}=0 \tag{8.4.1}$$

Like the other two modes of fracture, we are interested to find the stresses around the crack tip. Therefore, we introduce a stress function ϕ and define the shear stresses by

$$\tau_{xz}=\frac{\partial \phi}{\partial y} \qquad \tau_{yz}=-\frac{\partial \phi}{\partial x} \tag{8.4.2}$$

If we substitute Eq. (8.4.2) into Eq. (8.4.1) then we have:

$$\frac{\partial}{\partial x}\left(\frac{\partial \phi}{\partial y}\right)+\frac{\partial}{\partial y}\left(-\frac{\partial \phi}{\partial x}\right)=\frac{\partial^2 \phi}{\partial x \partial y}-\frac{\partial^2 \phi}{\partial x \partial y}=0$$

This proves that expressions in Eq. (8.4.2) are appropriate candidates for the stresses because it satisfies in equilibrium equation. If we write the Hook law for shear stress we have:

$$\gamma_{xz}=\frac{\tau_{xz}}{\mu} \qquad \gamma_{yz}=\frac{\tau_{yz}}{\mu} \tag{8.4.3}$$

In Eq. (8.4.3) μ is shear modulus If w is the displacement in z direction then γ_{xz} and γ_{yz} can be expressed by:

$$\gamma_{xz}=\frac{\partial w}{\partial x} \qquad \gamma_{yz}=\frac{\partial w}{\partial y} \tag{8.4.4}$$

According to Eq. (8.4.4) we can easily claim that the compatibility equation in mode III is:

$$\frac{\partial \gamma_{xz}}{\partial y} = \frac{\partial \gamma_{yz}}{\partial x} \tag{8.4.5}$$

This is obvious because if we substitute Eq. (8.4.4) into Eq. (8.4.5) we easily can obtain:

$$\frac{\partial \gamma_{xz}}{\partial y} = \frac{\partial \gamma_{yz}}{\partial x} = \frac{\partial^2 w}{\partial x \partial y}$$

If we substitute Eq. (8.4.2) into Eq. (8.4.3) and then substitute the result into Eq. (8.4.5) we have:

$$\frac{\partial}{\partial y}\left(\frac{1}{\mu}\frac{\partial \phi}{\partial y}\right) = -\frac{\partial}{\partial x}\left(\frac{1}{\mu}\frac{\partial \phi}{\partial x}\right) \tag{8.4.6}$$

Equation (8.4.6) can be written $\frac{1}{\mu}\left(\frac{\partial^2 \phi}{\partial y^2} + \frac{\partial^2 \phi}{\partial x^2}\right) = 0$ which can be reduced to:

$$\nabla^2 \phi = 0 \tag{8.4.7}$$

We conclude that stress function satisfies the Laplace equation. Therefore, in mode III fracture we are dealing with harmonic functions (not bi-harmonic). The stress formula from Eq. (8.4.7) can be obtained much easier. In Section 6.2, we mentioned that if we have a complex function $f(z)$, that its real part is ϕ and the imaginary part is ψ then we have:

$$f(z) = \phi(x, y) + i\psi(x, y) \tag{8.4.8}$$

In complex variable theory we can prove that both ϕ and ψ satisfy the Laplace equation so that we have:

$$\phi = \Re(f(z)) \quad \psi = \Im(f(z)) \tag{8.4.9}$$

In Eq. (8.4.9), $f(z)$ is analytic meaning that it can differentiated versus z. The question now is what is the stresses. To answer this we write the Cauchy Riemann compatibility:

$$\frac{\partial \phi}{\partial x}=\frac{\partial \psi}{\partial y} \qquad \frac{\partial \phi}{\partial y}=-\frac{\partial \psi}{\partial x} \tag{8.4.10}$$

If Eq. (8.4.10) is satisfied then, it does not matter from which path we differentiate we choose x path:

$$f^{'}(z)=\frac{df}{dz}=\frac{\partial \phi}{\partial x}+i\frac{\partial \psi}{\partial x} \tag{8.4.11}$$

If we substitute Eq. (8.4.10) into Eq. (8.4.11) then we have:

$$f^{'}(z)=\frac{\partial \phi}{\partial x}-i\frac{\partial \phi}{\partial y} \tag{8.4.12}$$

Now if we substitute the stress definition in Eq. (8.4.2) into Eq. (8.4.12) then we have:

$$f^{'}(z)=\tau_{yz}+i\,\tau_{xz} \qquad \tau_{yz}=\Re\left(f^{'}(z)\right) \qquad \tau_{xz}=\Im\left(f^{'}(z)\right) \tag{8.4.13}$$

From Eq. (8.4.13) we can see that they are much simpler than Koslov equations Eq. (6.3.14) to 6.3.16. From Eq. (8.4.13) stresses can be found easily. Only $f(z)$ should be an analytic function we propose this candidate for $f(z)$

$$f(z)=\tau\sqrt{z^2-a^2} \tag{8.4.14}$$

The above function is analytic all over the complex plane, and if we differentiate versus z:

$$f^{'}(z)=\frac{\tau z}{\sqrt{z^2-a^2}} \tag{8.4.15}$$

This is the same function that Westergaard proposed for mode III and can be justified because:

$$\lim_{z\to\infty} \frac{\tau z}{\sqrt{z^2 - a^2}} = \tau = \tau_{xz}\Big|_{x\to\infty} \tag{8.4.16}$$

The third reason for properness is that in the left tip $z = -a$ and in the right tip $z = a$ we see that stresses tend to infinity. This is also a physical fact for a crack tip. Therefore, Eqs. (8.4.14) and (8.4.15) can be justified. Then from Eqs. (8.4.13) and (8.4.15) we have:

$$\tau_{xz} = \tau\, \Im\left(\frac{z}{\sqrt{z^2 - a^2}} \right) \tag{8.4.17}$$

$$\tau_{yz} = \tau\, \Re\left(\frac{z}{\sqrt{z^2 - a^2}} \right) \tag{8.4.18}$$

Fortunately, in Eq. (8.3.12) and (8.3.18), we have calculated the above terms around the crack tip:

$$\Re\left(\frac{z}{\sqrt{z^2 - a^2}} \right) \cong \sqrt{\frac{a}{2r}} \cos\frac{\theta}{2} \qquad r << a \tag{8.4.19}$$

$$\Im\left(\frac{z}{\sqrt{z^2 - a^2}} \right) \cong -\sqrt{\frac{a}{2r}} \sin\frac{\theta}{2} \qquad r << a \tag{8.4.20}$$

By substituting Eqs. (8.4.19) and (8.4.20) into Eqs. (8.4.17) and (8.4.18) we calculate the stresses around the crack tip, that is, when $z \to a$ as follows:

$$\tau_{xz} = -\tau \sqrt{\frac{a}{2r}} \sin\frac{\theta}{2} \tag{8.4.21}$$

$$\tau_{xz} = \tau \sqrt{\frac{a}{2r}} \cos\frac{\theta}{2} \tag{8.4.22}$$

We define the stress intensity factor (SIF) in mode III, or K_{III} in a similar way to modes I and II, that is,

$$K_{III} = \tau \sqrt{\pi a} \tag{8.4.23}$$

If we substitute Eq. (8.4.23) into Eqs. (8.4.21) and (8.4.22) then we have:

$$\tau_{xz} = \frac{-K_{III}}{\sqrt{2\pi r}} \sin\frac{\theta}{2} \tag{8.4.24}$$

$$\tau_{yz} = \frac{K_{III}}{\sqrt{2\pi r}} \cos\frac{\theta}{2} \tag{8.4.25}$$

We multiply both sides of Eqs. (8.4.24) and (8.4.25) into $\sqrt{2\pi r}$ and also set $\theta = 0$ then we have:

$$K_{III} = \lim_{r\to 0}\left(\sqrt{2\pi r}\ \tau_{yz}\big|_{\theta=0}\right) \tag{8.4.26}$$

We can conclude that, SIF in mode III fracture can be obtained from Eq. (8.4.26). Therefore, shear stress in vicinity of crack should be studied and this can be done by FEM. It should be remembered that the photo-elasticity cannot measure out of plane stress and only numerical FEM type method to investigate on Mode III fracture.

KEYWORDS

- **anti-plane shear**
- **crack tip**
- **mode I fracture**
- **mode II fracture**
- **remote shear stress**
- **slipping mode**
- **stress intensity factor**
- **Westergaard function**

CHAPTER 9

ELASTIC-PLASTIC FRACTURE MECHANICS

CONTENTS

9.1 PLASTIC ZONE IN VICINITY OF CRACK TIP

In the previous chapters we discussed only about linear-elastic fracture mechanic (LEFM), in this theory the material is elastic and as we approach the crack tip, the stress value is going up and it can approach infinity. The LEFM is valid for brittle material like glass, and also some metals at low temperature.

The reason for validity of LEFM in these cases is because the plastic zone is very small and can be neglected. The rupture phenomenon is called brittle fracture in which fast fracture occurs. If the temperature in metals rises and if loading exceeds beyond a limit then the plastic zone becomes increasingly significant, and we cannot apply LEFM to study the fracture phenomena. In this case elastic-plastic fracture mechanic (EPFM) occurs.

One of the methods that can deal with plasticity around the crack, is finding an equivalent LEFM that can analyze EPFM. This technique is sometimes very useful, but occasionally the numerical FEM and experimental photo-elastic methods are also used. There is another branch of fracture mechanics that also covers EPFM and also fracture in nonlinear materials (does not obey Hook's law), which is called "nonlinear fracture mechanics" and such topic is beyond the scope of this book.

Before we make any connection between EPFM and ELFM, first we need to study the size of the plastic zone around the crack tip. Knowledge about the size of plastic region around the crack tip, enables, the best calibration of equivalent LEFM, to be applicable to EPFM. The most important question in EPFM is what is fracture toughness? This may sounds simple, but it helps the scientists to prepare appropriate samples to measure fracture toughness in EPFM by standard tests.

Knowledge about plastic yield criteria, is a key parameter in determination of the size of the plastic region. Yield criteria have been studied in many engineering courses in solid mechanics, including in the first part of this book. One of the well known criteria that is also explained in part one of the book is Von Mises yield criteria, which simply stating that if the total distortional strain energy, approaches the yield strain energy in uni-axial state of the stress, then we can claim that the material is plastically yielded. One way of writing the Von Mises law for yielding is expressing it in terms of the principal stresses into this form:

$$(\sigma_I - \sigma_{II})^2 + (\sigma_{II} - \sigma_{III})^2 + (\sigma_{III} - \sigma_I)^2 = 2\sigma_p^2 \qquad (9.1.1)$$

In Eq. (9.1.1) σ_p is the yield stress (not rupture stress) in simple uni-axial tensile test and σ_I, σ_{II} and σ_{III} are the principal stresses. These principal stresses are different from the ones that we described Eqs. (8.1.15), (8.1.16) and (8.1.17), we rewrite those again, that is,

$$\sigma_y = \frac{K_I}{\sqrt{2\pi r}} \cos\frac{\theta}{2}\left(1 - \sin\frac{\theta}{2}\sin\frac{3\theta}{2}\right) \qquad (9.1.2)$$

$$\sigma_x = \frac{K_I}{\sqrt{2\pi r}} \cos\frac{\theta}{2}\left(1 + \sin\frac{\theta}{2}\sin\frac{3\theta}{2}\right) \qquad (9.1.3)$$

$$\tau_{xy} = \frac{K_I}{\sqrt{2\pi r}} \sin\frac{\theta}{2}\cos\frac{\theta}{2}\cos\frac{3\theta}{2} \qquad (9.1.4)$$

It should be reminded that there is not the term $-\sigma$ in Eq. (9.1.3), as it appears in Eq. (8.4.16), the reason is that the remote stress σ is very small compared to the stresses at crack tip, or we can say $\sigma << \frac{K_I}{\sqrt{2\pi r}} \quad r << a$, therefore, ignoring the term $-\sigma$ in Eq. (9.1.3) can be justified. Now we have to find out what is the relationship between σ_y, σ_x and τ_{xy} with the principal stresses σ_I, σ_{II} and σ_{III}? The answer to this question is given in elementary solid mechanics by the Mohr circle, in which the relationship between the principal stresses and the general stresses can be found from the following formulas (see Fig. 9.1).

$$\sigma_I = \frac{\sigma_x + \sigma_y}{2} + \sqrt{\left(\frac{\sigma_x - \sigma_y}{2}\right)^2 + \tau_{xy}^2} \tag{9.1.5}$$

$$\sigma_I = \frac{\sigma_x + \sigma_y}{2} - \sqrt{\left(\frac{\sigma_x - \sigma_y}{2}\right)^2 + \tau_{xy}^2} \tag{9.1.6}$$

In the left side the Mohr circle is drawn which clarifies the structure of the formulas in Eqs. (9.1.5) and (9.1.6). If we add up Eqs. (9.1.2) and (9.1.3) halve the result then we have:

$$\frac{\sigma_x + \sigma_y}{2} = \frac{K_I}{\sqrt{2\pi r}} \cos\frac{\theta}{2} \tag{9.1.7}$$

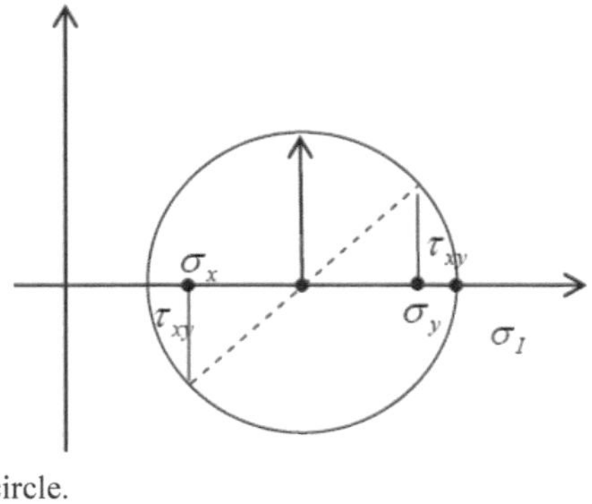

FIGURE 9.1 Mohr circle.

If we take away Eq. (9.1.2) from Eq. (9.1.3) and halve the result then we have:

$$\frac{\sigma_x - \sigma_y}{2} = \frac{K_I}{\sqrt{2\pi r}} \cos\frac{\theta}{2} \sin\frac{\theta}{2} \sin\frac{3\theta}{2} \tag{9.1.8}$$

If we substitute Eqs. (9.1.7), (9.1.8) and (9.1.4) into Eqs. (9.1.5) and (9.1.6) then we have:

$$\sigma_I = \frac{K_I}{\sqrt{2\pi r}} \cos\frac{\theta}{2} + \frac{K_I}{\sqrt{2\pi r}} \sqrt{\left(\cos\frac{\theta}{2} \sin\frac{\theta}{2} \sin\frac{3\theta}{2}\right)^2 + \left(\sin\frac{\theta}{2} \cos\frac{\theta}{2} \cos\frac{3\theta}{2}\right)^2}$$

$$\sigma_{II} = \frac{K_I}{\sqrt{2\pi r}} \cos\frac{\theta}{2} - \frac{K_I}{\sqrt{2\pi r}} \sqrt{\left(\cos\frac{\theta}{2} \sin\frac{\theta}{2} \sin\frac{3\theta}{2}\right)^2 + \left(\sin\frac{\theta}{2} \cos\frac{\theta}{2} \cos\frac{3\theta}{2}\right)^2}$$

It is obvious that the above equations can be simplified into:

$$\sigma_I = \frac{K_I}{\sqrt{2\pi r}} \cos\frac{\theta}{2} + \frac{K_I}{\sqrt{2\pi r}} \cos\frac{\theta}{2} \sin\frac{\theta}{2} \sqrt{\underbrace{\sin^2\frac{3\theta}{2} + \cos^2\frac{3\theta}{2}}_{1}}$$

$$\sigma_{II} = \frac{K_I}{\sqrt{2\pi r}} \cos\frac{\theta}{2} - \frac{K_I}{\sqrt{2\pi r}} \cos\frac{\theta}{2} \sin\frac{\theta}{2} \sqrt{\underbrace{\sin^2\frac{3\theta}{2} + \cos^2\frac{3\theta}{2}}_{1}}$$

The final simplified form is:

$$\sigma_I = \frac{K_I}{\sqrt{2\pi r}} \cos\frac{\theta}{2}\left(1 + \sin\frac{\theta}{2}\right) \tag{9.1.9}$$

$$\sigma_{II} = \frac{K_I}{\sqrt{2\pi r}} \cos\frac{\theta}{2}\left(1 - \sin\frac{\theta}{2}\right) \tag{9.1.10}$$

With plane stress state we have $\sigma_{III} = 0$ and the Eq. (9.1.1) can be rewritten into this form:

$$\sigma_I^2 + \sigma_{II}^2 - \sigma_I \sigma_{II} = \sigma_p^2 \tag{9.1.11}$$

The above equation is also well known an shown in many solid mechanic books. If we substitute Eqs. (9.1.9) and (9.1.10) into Eq. (9.1.11) then we have:

$$\left(\frac{K_I}{\sqrt{2\pi r}}\right)^2 \begin{bmatrix} \cos^2\frac{\theta}{2}\left(1+\sin\frac{\theta}{2}\right)^2 + \cos^2\frac{\theta}{2}\left(1-\sin\frac{\theta}{2}\right)^2 + \\ \cos^2\frac{\theta}{2}\left(1+\sin\frac{\theta}{2}\right)\left(1-\sin\frac{\theta}{2}\right) \end{bmatrix} = \sigma_p^2$$

Exercise 9.1.1: *Simplify the above equation*

Solution: By factorization and simplification we have the following form: $\left(\frac{K_I}{\sqrt{2\pi r}}\cos\frac{\theta}{2}\right)^2\left[\left(1+\sin\frac{\theta}{2}\right)^2+\left(1-\sin\frac{\theta}{2}\right)^2+\left(1-\sin^2\frac{\theta}{2}\right)\right]=\sigma_p^2$ and will change to:

$\left(\frac{K_I}{\sqrt{2\pi r}}\right)^2\cos^2\frac{\theta}{2}\left(2+2\sin^2\frac{\theta}{2}-\cos^2\frac{\theta}{2}\right)=\sigma_p^2$, then from the trigonometry the form will change to: $\left(\frac{K_I}{\sqrt{2\pi r}}\right)^2\cos^2\frac{\theta}{2}\left(1+3\sin^2\frac{\theta}{2}\right)=\sigma_p^2$ by multiplying $\cos^2\frac{\theta}{2}$ to the bracket:

$\left(\frac{K_I}{\sqrt{2\pi r}}\right)^2\left(\cos^2\frac{\theta}{2}+3\sin^2\frac{\theta}{2}\cos^2\frac{\theta}{2}\right)=\sigma_p^2$ and again by using full arc formulas:

$$\left(\frac{K_I}{\sqrt{2\pi r}}\right)^2\left(\frac{1+\cos\theta}{2}+\frac{3}{4}\sin^2\theta\right)=\sigma_p^2$$

The r that can be found from the above formula is specified by r_p or radius of the plastic zone in vicinity of the crack. The above equation then will be:

$$\frac{K_I}{2\pi r_p}\left(1+\cos\theta+\frac{3}{2}\sin^2\theta\right)=2\sigma_p^2 \tag{9.1.12}$$

Or:

$$r_p\left(K_I,\theta\right)=\frac{\left(1+\cos\theta+\frac{3}{2}\sin^2\theta\right)K_I^2}{4\pi\sigma_p^2} \tag{9.1.13}$$

The advantage of the Eq. (9.1.13) is that it can be drawn on a polar diagram so that distribution of the plastic zone radius around the crack tip can be displayed. For the plane strain case $\sigma_{III}\neq 0$ and we need to substitute $\sigma_{III}=\upsilon\left(\sigma_I+\sigma_{II}\right)$ in the Eq. (9.1.1) then we have:

$$\left(\sigma_I-\sigma_{II}\right)^2+\left(\left(1-\upsilon\right)\sigma_{II}-\upsilon\sigma_I\right)^2+\left(\left(1-\upsilon\right)\sigma_I-\upsilon\sigma_{II}\right)^2=2\sigma_p^2$$

Exercise 9.1.2: *Simplify the above equation*

Solution: We can factorize the terms σ_I^2, σ_{II}^2 and also $\sigma_I\sigma_{II}$ then we have:
$\left(1+\left(1-\upsilon\right)^2+\upsilon^2\right)\sigma_I^2+\left(1+\left(1-\upsilon\right)^2+\upsilon^2\right)\sigma_{II}^2-4\upsilon\left(1-\upsilon\right)\sigma_I\sigma_{II}-2\sigma_I\sigma_{II}=2\sigma_p^2$
and by factorizing we have, $2\left(\sigma_I^2+\sigma_{II}^2\right)\left(1+\upsilon^2-\upsilon\right)+4\sigma_I\sigma_{II}\left(1+\upsilon^2-\upsilon\right)-6\sigma_I\sigma_{II}=2\sigma_p^2$ and then further simplification we have:

$$\left(\sigma_I+\sigma_{II}\right)^2\left(1+\upsilon^2-\upsilon\right)-3\sigma_I\sigma_{II}=\sigma_p^2 \tag{9.1.14}$$

If we add up Eqs. (9.1.9) to (9.1.10) and the square both sides we have:

$$\left(\sigma_I+\sigma_{II}\right)^2=4\left(\frac{K_I}{\sqrt{2\pi r}}\right)^2\cos^2\frac{\theta}{2} \tag{9.1.15}$$

If we multiply Eqs. (9.1.9) to (9.1.10) then we have:

$$\sigma_I\sigma_{II}=\left(\frac{K_I}{\sqrt{2\pi r}}\right)^2\cos^4\frac{\theta}{2} \tag{9.1.16}$$

If we substitute Eqs. (9.1.16) and (9.1.15) into Eq. (9.1.14) then we have:

$$\left(\frac{K_I}{\sqrt{2\pi r}}\right)^2 \cos^2\frac{\theta}{2}\left(4\left(1+\upsilon^2-\upsilon\right)-3\cos^2\frac{\theta}{2}\right)=\sigma_p^2$$

***Exercise* 9.1.3:** *Modify the above equation into an appropriate form*

Solution: By trigonometry and multiplication the above equation changes to: $\left(\frac{K_I}{\sqrt{2\pi r}}\right)^2 \cos^2\frac{\theta}{2}\left(4+4\upsilon^2-4\upsilon-3\left(1-\sin^2\frac{\theta}{2}\right)\right)=\sigma_p^2$, by considering $\cos^2\frac{\theta}{2}=\frac{1+\cos\theta}{2}$
and also knowing that $1+4\upsilon^2-4\upsilon=\left(1-2\upsilon\right)^2$, and then multiply both sides to 2, we have:

$\left(\frac{K_I}{\sqrt{2\pi r}}\right)^2\left(\left(1-2\upsilon\right)^2\left(1+\cos\theta\right)+6\sin^2\frac{\theta}{2}\cos^2\frac{\theta}{2}\right)=2\sigma_p^2$, from trigonometry we have

$6\sin^2\frac{\theta}{2}\cos^2\frac{\theta}{2}=\frac{3}{2}\sin^2\theta$ and if we substitute into the above we have:

$$\left(\frac{K_I}{\sqrt{2\pi r}}\right)^2\left(\left(1-2\upsilon\right)^2\left(1+\cos\theta\right)+\frac{3}{2}\sin^2\theta\right)=2\sigma_p^2 \qquad (9.1.17)$$

The r that can be found from the above formula is specified by r_p or radius of the plastic zone in vicinity of the crack. The above equation then will be:

$$r_p\left(K_I,\theta\right)=\frac{\left(\left(1-2\upsilon\right)^2\left(1+\cos\theta\right)+\frac{3}{2}\sin^2\theta\right)K_I^2}{4\pi\sigma_p^2} \qquad (9.1.18)$$

It is very interesting to compare Eq. (9.1.13) for plane stress case and Eq. (9.1.18) for plane strain case. Both have the same form only in plane strain the coefficient of $\left(1+\cos\theta\right)$ is contracted by the factor $\left(1-2\upsilon\right)^2 \ll 1$, and then we can conclude that the polar diagram that shows the size of r_p in plane strain state is much smaller and this fact is shown in Fig. 9.2, in which the dashed area is the plastic zone for plain strain case.

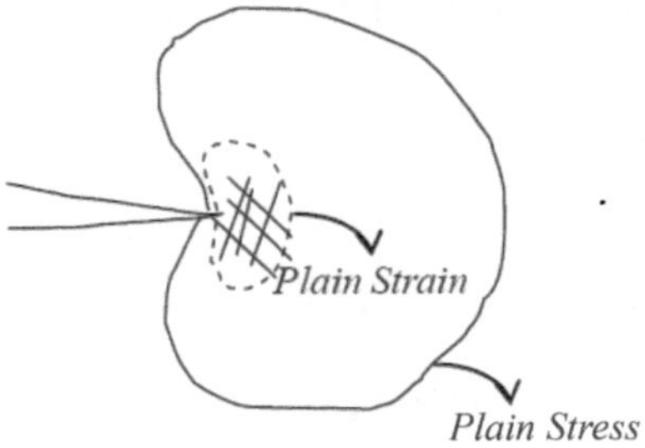

FIGURE 9.2 Plastic zone.

We can conclude that, since the plastic region in plane strain state is very small (see Fig. 9.2), for measuring the fracture toughness by the experiment it is better to conduct the test sample in plane strain case. This means that the thickness of the sample should be much higher, so that plane strain can be implemented. Therefore, fracture toughness test are expensive needs powerful machines to apply higher load to cause-to-cause fast fracture, thereby measuring the fracture toughness. In the next section we discuss how the Irwin provided some instructions to conduct tests in EPFM.

9.2 IRWIN ANALYSIS OF EPFM

Irwin is the founder of the modern fracture mechanics. The first question for him was that in ductile fracture (Fig. 9.3) there is plastic zone around the crack tip, can we still use LEFM to analyze this? The answer is not easy because as we saw in the last section that radius of plastic zone is function of, θ, that is, $r_p\left(K_I,\theta\right)$.

We saw that the size for plain strain case is much smaller. Irwin proposed an approximate method in which he calculated $\lim_{\theta\to 0}\sigma_y$ and we

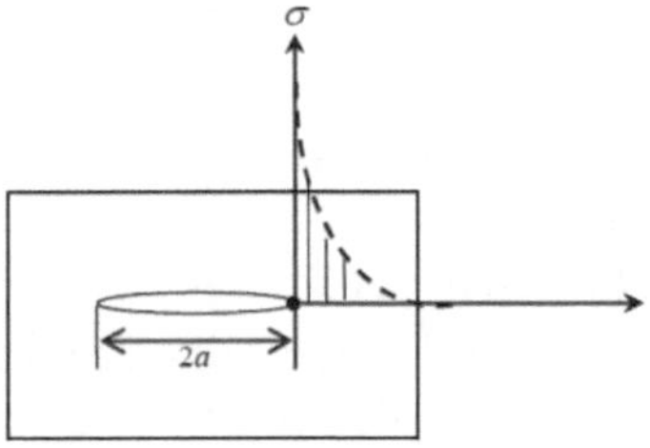

FIGURE 9.3 Stress distribution.

write it again, that is, $\lim_{\theta\to 0}\sigma_y = \frac{K_I}{\sqrt{2\pi r}}\lim_{\theta\to 0}\cos\frac{\theta}{2}\left(1-\sin\frac{\theta}{2}\sin\frac{3\theta}{2}\right)$ which can be simplified to:

$$\sigma_y\big|_{\theta=0} = \frac{K_I}{\sqrt{2\pi r}} \tag{9.2.1}$$

Then he argued that the stress cannot exceed σ_p or the yield stress, and therefore in Eq. (9.2.1) we substitute $r = r_y$ to find out the corresponding radius for $\sigma_y = \sigma_p$, that is,

$\sigma_p = \frac{K_I}{\sqrt{2\pi r_y}} \Rightarrow \sigma_p^2 = \frac{K_I^2}{2\pi r_y}$ which results:

$$r_y = \frac{1}{2\pi}\left(\frac{K_I}{\sigma_p}\right)^2 \tag{9.2.2}$$

We rewrite the Eq. (9.1.14) again, that is,

$$(\sigma_I + \sigma_{II})^2\left(1+\upsilon^2-\upsilon\right)-3\sigma_I\sigma_{II} = \sigma_p^2 \tag{9.2.3}$$

The principal stresses are given by Eqs. (9.1.9) and (9.1.10), and if we set $\theta = 0$, then $\sigma' = \sigma_I\big|_{\theta=0} = \sigma_{II}\big|_{\theta=0} = \frac{K_I}{\sqrt{2\pi r}}$, for plane stress case we need to set $\upsilon = 0$ in Eq. (9.2.3) and if we set $\sigma' = \sigma_I\big|_{\theta=0} = \sigma_{II}\big|_{\theta=0}$ as well, the equation will change to: $\left(2\sigma'\right)^2 - 3\sigma'^2 = \sigma_p^2$, or $\sigma' = \sigma_p$

which means that to reach the yield condition around crack $\sigma' \to \sigma_p$. It is obvious that for plain strain condition yield condition will be different. It is obvious that Eq. (9.2.2) can be valid for Von Mises yield condition. According to Fig. 9.4, in the first instance it seems that in zone with radius r_y, the stress is uniform with value σ_p and after that stress is declining. Irwin showed that, the above assumption cannot be correct.

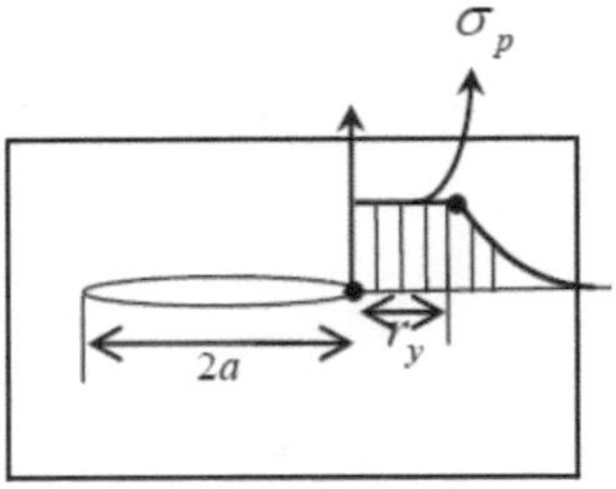

FIGURE 9.4 Initial assumption.

He says that, for using LEFM to analyse EPFM, initially it is necessary to assume that in vicinity $0 < r < r_y$ the material is hypothetically elastic (not really) then next step is to calculate how much force is necessary to keep the static equilibrium. The force per unit of thickness is $\frac{F}{B}$ and can be calculated as follows:

$\frac{F}{B} = \int_0^{r_y} \sigma_y(r)\,dr = \int_0^{r_y} \frac{K_I}{\sqrt{2\pi r}}\,dr = \frac{K_I}{\sqrt{2\pi}}\left(2r^{1/2}\right)_0^{r_y}$ which after substitution of upper bounds:

$$\frac{F}{B} = \frac{K_I}{\sqrt{2\pi}} r_y^{\ 1/2} \tag{9.2.4}$$

Now if we substitute r_y from Eq. (9.2.2) into Eq. (9.2.4) then we have:

$$\frac{F}{B} = \frac{K_I}{\sqrt{2\pi}}\left(\frac{1}{2\pi}\left(\frac{K_I}{\sigma_p}\right)^2\right)^{1/2} = \frac{K_I^2}{\pi\sigma_p} \tag{9.2.5}$$

Then Irwin said that radius of the plastic zone is r_p may not be r_y but some value that can produce same $\frac{F}{B}$ in region $0 < r < r_p$ with a constant yield stress σ_p, that is,

$$\frac{F}{B} = \sigma_p r_p \tag{9.2.6}$$

If we compare Eqs. (9.2.5) and (9.2.6) then we have:

$$\sigma_p r_p = \frac{K_I^2}{\pi \sigma_p} \Rightarrow r_p = \frac{K_I^2}{\pi \sigma_p^2} \tag{9.2.7}$$

If we compare Eqs. (9.2.2) and (9.2.7) then:

$$r_p = 2r_y \tag{9.2.8}$$

Equation (9.2.8) means that the plastic radius is twice the value that we initially suggested. Thereafter Irwin said that if we want to use LEFM to analyze the problems in EPFM, we need to follow these procedures:

i. Displace the crack tip by r_y, that is, assume a hypothetical crack with the total length $2\left(a + r_y\right)$ and modify the SIF based on this equation:

$$K_{\mathrm{Im}} = \sigma\sqrt{\pi\left(a + r_y\right)} \tag{9.2.9}$$

ii. In the new virtual crack consider a region $0 < r < r_y$ in which the plastic stress σ_p is constant (see Fig. 9.5) and this can simulate the real situation in which total region is $0 < r < r_p$ in the real crack, but half of that region faces opening in the virtual crack.

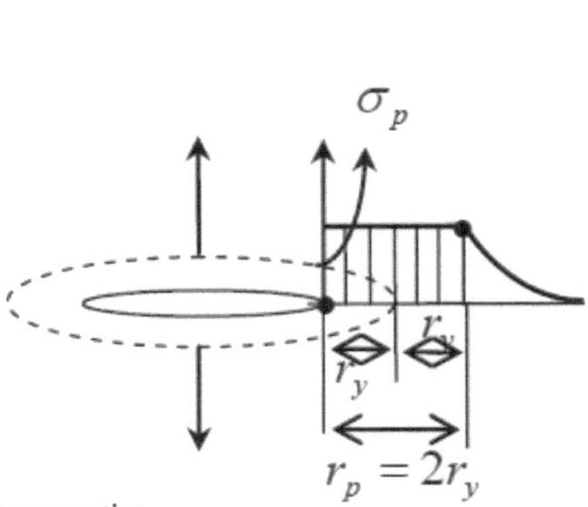

FIGURE 9.5 Irwin's suggestion.

According to Irwin the above statement guarantees that not only the static equilibrium can be hold, but also we can use the Eq. (9.2.9) for the modified SIF and implement LEFM for analyzing EPFM.

As we mentioned in the virtual crack half of the real plastic region is facing the opening and the amount of this opening has been studied before in Chapter 7, by Eq. (7.1.12) and now we rewrite it for the virtual crack into the following form:

$$v=\left(\frac{\kappa+1}{4\mu}\right)\sigma\sqrt{\left(a+r_y\right)^2-x^2} \tag{9.2.10}$$

For plane stress state we have, $\mu=\dfrac{E}{2(1+\upsilon)}$ and $\kappa=\dfrac{3-\upsilon}{1+\upsilon}$ by substituting it in Eq. (9.2.10) we have:

$$v=\frac{2\sigma}{E}\sqrt{\left(a+r_y\right)^2-x^2} \tag{9.2.11}$$

In Eq. (9.2.11) if we substitute $x=a$, then we have:

$$v_{x=a}=\frac{2\sigma}{E}\sqrt{\left(a+r_y\right)^2-a^2}=\frac{2\sigma}{E}\sqrt{r_y^2+2ar_y} \tag{9.2.12}$$

Irwin emphasized that his suggestions are valid only if the size of the plastic zone is very small, that is, $r_y \ll a$, therefore in Eq. (9.2.12) we can ignore the term r_y^2 and then the full crack tip opening displacement (CTOD) can be determined by:

$$\delta=2v_{x=a}\cong\frac{4\sigma}{E}\sqrt{2ar_y} \tag{9.2.13}$$

In Eq. (9.2.13) δ, is a hypothetical value because there is not any opening in the crack tip for real crack. However, in the virtual crack that suggested by Irwin δ does exist and later in UK they found out that δ (CTOD) is an important factor in fracture mechanics. If we substitute r_y from Eq. (9.2.2) into Eq. (9.2.13), then we have:

$\delta\cong\dfrac{2\sigma}{E}\sqrt{2a\dfrac{1}{2\pi}\left(\dfrac{K_I}{\sigma_p}\right)^2}\cong\dfrac{4\sigma}{E}\dfrac{K_I}{\pi\sigma_p}$, but we have $K_I=\sigma\sqrt{\pi a}$ from LEFM which is different from K_{Im} in EPFM defined in Eq. (9.2.9) by Irwin. Then Eq. (9.2.13) can be fully expressed in terms of K_I into this form:

$$\delta \cong \frac{4K_I^2}{\pi E \sigma_p} \qquad (9.2.14)$$

It is obvious that $K_{\mathrm{Im}} > K_I$ confirming that plasticity can increase fracture toughness, and they way of connecting σ_p or yield stress to fracture toughness in via δ (CTOD) in Eq. (9.2.14). This is the only reason of δ being an important parameter. If the applied load (remote stress σ) increases then $K_I \to K_{IC}$ or the fracture toughness and then $\delta \to \delta_C$ or the critical CTOD so that Eq. (9.2.14) will be:

$$\delta_C \cong \frac{4K_{IC}^2}{\pi E \sigma_p} \qquad (9.2.15)$$

In Chapter 7, we have discussed about critical energy release rate or G_C which is important in fracture mechanics and is related to fracture toughness by Eq. (7.2.15), which for plane stress case is:

$$G_C = \frac{K_{IC}^2}{E} \qquad (9.2.16)$$

Substituting Eq. (9.2.16) into Eq. (9.2.15) results:

$$G_C \cong \left(\frac{\pi}{4}\right)\sigma_p \, \delta_C \qquad (9.2.17)$$

Importance of critical CTOD δ_C can be revealed by Eq. (9.2.17), which provides a criteria for finding G_C in EPFM.

For plane strain situation, the conclusions are different. for plane strain case we need to set $\upsilon \neq 0$ in (9.2.3) and if we set $\upsilon = 1/3$ and also considering that $\sigma' = \sigma_I|_{\theta=0} = \sigma_{II}|_{\theta=0} = \dfrac{K_I}{\sqrt{2\pi r}}$, the equation will Eq. (9.2.3) change to: $(7/9)(2\sigma')^2 - 3\sigma'^2 = \sigma_p^2$, or $\sigma' = 3\sigma_p$ which means that to reach the yield condition around crack $\sigma' \to 3\sigma_p$ for $\upsilon = 1/3$.

In previous section, we said that always use plane strain state for conducting the tests to measure the fracture toughness K_{IC}, because in plane strain case, plastic zone is small and LEFM can be implemented.

When Irwin first conducted tests to measure the fracture toughness, he used a cracked rod. However, for plane strain case, the thickness of the sample should be big, otherwise the Eq. (9.2.3) is not applicable. Therefore, he said that for a cracked rod $\sigma' = \sqrt{3}\,\sigma_p$, the proof is not available in public domain. He insisted that for doing any fracture toughness test, it is necessary that crack length a, thickness B and the width W should be much bigger than plastic radius r_p and then he proposed the following relation:

$$a, W, B \geq 50 r_y \tag{9.2.18}$$

If in Eq. (9.2.2) we change σ_p to $\sqrt{3}\,\sigma_p$ then:

$$r_y = \frac{1}{2\pi}\left(\frac{K_{IC}}{\sqrt{3}\,\sigma_p}\right)^2 \tag{9.2.19}$$

If we substitute Eq. (9.2.19) into Eq. (9.2.18) then we have:

$a, W, B \geq \frac{50}{6\pi}\left(\frac{K_{IC}}{\sigma_p}\right)^2$ and since $\frac{50}{6\pi} \cong 2.5$, then Eq. (9.2.18) will change to a formula in standard, that is,

$$a, W, B \geq 2.5\left(\frac{K_{IC}}{\sigma_p}\right)^2 \tag{9.2.20}$$

In ductile material K_{IC} is very big, and according to Eq. (9.2.20) not only a big specimen should be made, the rack size should also be big. Generating such big cracks in large samples needs a fatigue machine to run to generate a big crack in large sample and after that the fracture toughness test can be carried out. Therefore, experiments in fracture mechanics are very expensive and sometimes-numerical simulations are preferred.

9.3 WESTERGAARD FUNCTION FOR CONCENTRATED FORCE ON CRACK FACE

In this section, we study about the stress function, when a concentrated force (Fig. 9.6) is applied to the crack face and then determine SIF for that.

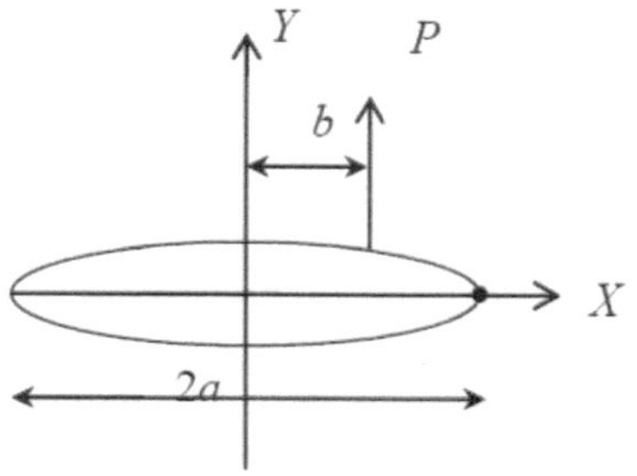

FIGURE 9.6 Concentrated force.

The purpose is further investigation into EPFM and analyze the Dugdale method. It is obvious that Westergaard functions in this case have simple form, that is, at $z = a$ and also $z = -a$ therefore $\frac{1}{\sqrt{z^2 - a^2}}$ will be part of that function. Moreover, in point $z = b$, the stress is Moreover, in point $z = b$, the stress is singular and this necessitates the factor $\frac{1}{z - b}$ as well. We showed that Westergaard function has the simplest form and therefore it can take this form:

$$Z_I(z) = \frac{C}{(z-b)\sqrt{z^2 - a^2}} \tag{9.3.1}$$

In Eq. (9.3.1), C is a coefficient that should be determined. If rewrite Eq. (7.3.12) over the x axis or $y = 0$ again we have:

$$\sigma_y = \Re(Z_I(z)) + A \tag{9.3.2}$$

In Eq. (9.3.2), $A = 0$, since we have concentrated force and the remote stress $\sigma = 0$, therefore we have:

$$\sigma_y\big|_{y=0} = \frac{C}{(x-b)\sqrt{x^2 - a^2}} \tag{9.3.3}$$

Then we can write:

$$P = \int_{-\infty}^{a} \sigma_y\big|_{y=0} dx + \int_{a}^{\infty} \sigma_y\big|_{y=0} dx = \int_{-\infty}^{a} \frac{C\,dx}{(x-b)\sqrt{x^2 - a^2}} + \int_{a}^{\infty} \frac{C\,dx}{(x-b)\sqrt{x^2 - a^2}}$$

(9.3.4)

Equation (9.3.4) is static equilibrium equation and to perform the above integrals we set $x-b=u$, and then we have $x^2-a^2=u^2+2bu+b^2-a^2$, and then (9.3.4) becomes:

$$P=\int_{-\infty}^{-a-b}\frac{C\,du}{u\sqrt{u^2+2bu+b^2-a^2}}+\int_{a-b}^{\infty}\frac{C\,du}{u\sqrt{u^2+2bu+b^2-a^2}} \tag{9.3.5}$$

Exercise 9.3.1: *Evaluate the integrals in* Eq. (9.3.5)

Solution: By using symbolic math toolbox in MATLAB we have:

$$F(u)=\int\frac{du}{u\sqrt{u^2+2bu+b^2-a^2}}=\frac{-\ln\left(b-\dfrac{b^2-a^2}{u}+\dfrac{\sqrt{b^2-a^2}}{u}\sqrt{u^2+2bu+b^2-a^2}\right)}{\sqrt{b^2-a^2}} \tag{9.3.6}$$

Considering Eq. (9.3.6) we can write Eq. (9.3.5) into this form:

$$P=\left(F(-a-b)\underbrace{-F(-\infty)+F(\infty)}_{0}-F(a-b)\right) \tag{9.3.7}$$

We have: $F(\pm\infty)=\lim_{u\to\infty}\dfrac{-\ln\left(b-\dfrac{b^2-a^2}{u}+\sqrt{b^2-a^2}\sqrt{1+2\dfrac{b}{u}+\dfrac{b^2-a^2}{u^2}}\right)}{\sqrt{b^2-a^2}}$

which results:

$$F(\infty)=F(-\infty)=\frac{-\ln\left(b+\sqrt{b^2-a^2}\right)}{\sqrt{b^2-a^2}} \tag{9.3.8}$$

$$F(-a-b)=$$
$$\frac{-\ln\left(b-\frac{b^2-a^2}{-a-b}+\frac{\sqrt{b^2-a^2}}{-a-b}\sqrt{\underbrace{(-a-b)^2+2b(-a-b)+b^2-a^2}_{0}}\right)}{\sqrt{b^2-a^2}}$$ which results:

$$F(-a-b)=\frac{-\ln(b+a-b)}{\sqrt{b^2-a^2}}=\frac{-\ln a}{\sqrt{b^2-a^2}} \tag{9.3.9}$$

$$F(a-b)=\frac{-\ln\left(b-\frac{b^2-a^2}{a-b}+\frac{\sqrt{b^2-a^2}}{a-b}\sqrt{\underbrace{(a-b)^2+2b(a-b)+b^2-a^2}_{0}}\right)}{\sqrt{b^2-a^2}}$$

which results:

$$F(a-b)=\frac{-\ln(b-a-b)}{\sqrt{b^2-a^2}}=\frac{-\ln(-a)}{\sqrt{b^2-a^2}} \tag{9.3.10}$$

We substitute Eqs. (9.3.8), (9.3.9) and (9.3.10) into Eq. (9.3.7), then we have:

$$P=\left(\frac{-\ln a}{\sqrt{b^2-a^2}}+\frac{\ln(-a)}{\sqrt{b^2-a^2}}\right)C=\frac{C\ln(-1)}{\sqrt{b^2-a^2}} \tag{9.3.11}$$

However we have $\ln(-1)=\ln\left(e^{i\pi}\right)=i\pi,$ also we have $\sqrt{b^2-a^2}$ $i\sqrt{a^2-b^2}$ substituting in Eq. (9.3.11) provides:

$P=\dfrac{iC\pi}{i\sqrt{a^2-b^2}}\Rightarrow C=\dfrac{P\sqrt{a^2-b^2}}{\pi}$, then (9.3.11) will change to:

$$Z_{Ir}(z)=\frac{P\sqrt{a^2-b^2}}{\pi(z-b)\sqrt{z^2-a^2}} \tag{9.3.12}$$

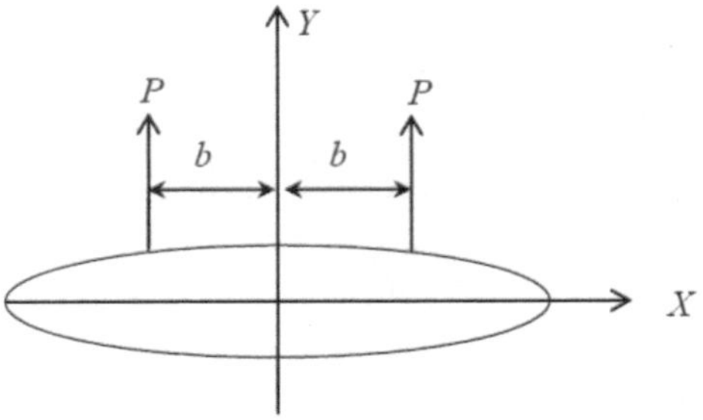

FIGURE 9.7 Pair of forces.

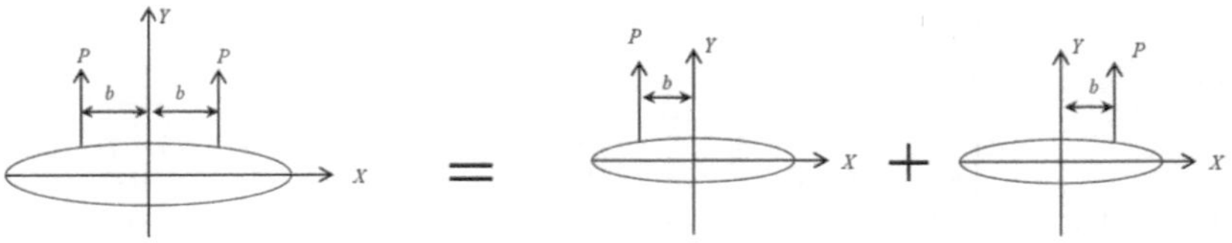

FIGURE 9.8 Superposition principle.

The index r in (9.3.12) means that, concentrated force is applied in the right side of the crack. Now consider a pair of forces (one in right the other in left) shown in Fig. 9.7, both with distance b from middle of the crack, in this case we use superposition principle according to Fig. 9.8., and we are allowed to do so, because the governing equations (bi-harmonic) are linear. We need to introduce $Z_{II}(z)$ as well which is:

$$Z_{II}(z) = \frac{P\sqrt{a^2 - b^2}}{\pi(z+b)\sqrt{z^2 - a^2}} \tag{9.3.13a}$$

According to the superposition principle $Z_I(z)$ for the pair of force is summation of $Z_{II}(z)$ and $Z_{Ir}(z)$, that is, $Z_I(z) = Z_{Ir}(z) + Z_{II}(z)$ $= \dfrac{P\sqrt{a^2 - b^2}}{\pi\sqrt{z^2 - a^2}}\left(\dfrac{1}{z-b} + \dfrac{1}{z+b}\right)$ which can be simplified to:

$$Z_I(z) = \frac{2Pz\sqrt{a^2 - b^2}}{\pi(z^2 - b^2)\sqrt{z^2 - a^2}} \tag{9.3.13b}$$

In order to find the SIF in this case, we need to return to Chapter 8, and rewrite, Eq. (8.1.6), that is,

$$K_I = \sqrt{2\pi}\, \lim_{z \to a} \sqrt{z-a}\, Z_I(z) \tag{8.1.6}$$

If we substitute Eq. (9.3.13b) in Eq. (8.1.6) then we have:

$K_I = \sqrt{2\pi}\, \lim_{z \to a} \left(\sqrt{z-a} \dfrac{2Pz\sqrt{a^2-b^2}}{\pi\left(z^2-b^2\right)\sqrt{z^2-a^2}} \right)$, but it can be simplified since:

$\lim_{z \to a} \left(\dfrac{\sqrt{z-a}}{\sqrt{z^2-a^2}} \right) = \lim_{z \to a} \left(\dfrac{1}{\sqrt{z+a}} \right) = \dfrac{1}{\sqrt{2a}}$ and if we substitute this SIF will be:

$$K_I = \frac{2P\sqrt{\pi a}}{\pi\sqrt{a^2-b^2}} \tag{9.3.14}$$

To describe the Dugdale method, in the next section we need to have a formula that shows the SIF when in a crack with the length $2(a+\rho)$, a yield stress σ_p is applied in a region between a and $a+\rho$ shown in Fig. 9.9, then Eq. (9.3.14) will help to find out the SIF for such case, it is only necessary to replace $a \Leftrightarrow a+\rho$ and $P \Leftrightarrow \sigma_p db$ also $K_I \Leftrightarrow dK_I$ in Eq. (9.3.14), the Eq. (9.3.14) will be:

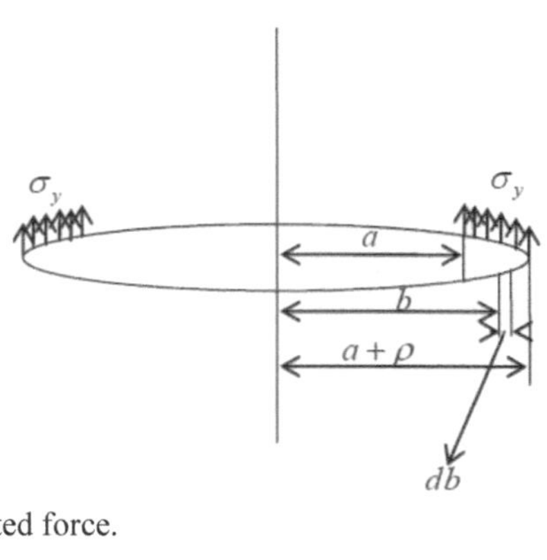

FIGURE 9.9 Distributed force.

$$dK_I = \frac{2\sigma_p \sqrt{\pi(a+\rho)}\, db}{\pi \sqrt{(a+\rho)^2 - b^2}} \tag{9.3.15}$$

If we integrate from Eq. (9.3.15), the SIF as a result of distributed stress in σ_p the region $a < b < a+\rho$ in each side of the crack.

$$K_I = \int_a^{a+\rho} \frac{2\sigma_p \sqrt{\pi(a+\rho)}}{\pi \sqrt{(a+\rho)^2 - b^2}}\, db = \frac{2\sigma_p \sqrt{\pi(a+\rho)}}{\pi} \int_a^{a+\rho} \frac{db}{\sqrt{(a+\rho)^2 - b^2}} \tag{9.3.16}$$

Exercise 9.3.2: *Evaluate the integral* $\int_a^{a+\rho} \frac{db}{\sqrt{(a+\rho)^2 - b^2}}$

Solution: We can use the symbolic math toolbox in MATLAB and integrate versus the variable b with lower bound a and the upper bound $a+\rho$, the following script file is needed:

```
syms b a ro
A=1/sqrt((a+ro)^2b^2)
B=int(A, b, a, (a+ro))
```

The result which is given by MATLAB is:

$$\int_a^{a+\rho} \frac{db}{\sqrt{(a+\rho)^2 - b^2}} = \frac{\pi}{2} - \sin^{-1}\left(\frac{a}{a+\rho}\right) \tag{9.3.17}$$

We assume that the arc in Eq. (9.3.17) is $\frac{\pi}{2} - \sin^{-1}\left(\frac{a}{a+\rho}\right) = \beta$ and if we take cosine of this expression then we have, $\cos\beta = \cos\left(\frac{\pi}{2} - \sin^{-1}\left(\frac{a}{a+\rho}\right)\right) = \sin\left(\sin^{-1}\left(\frac{a}{a+\rho}\right)\right)$ then it is obvious that $\cos\beta = \frac{a}{a+\rho}$ and results $\beta = \cos^{-1}\left(\frac{a}{a+\rho}\right)$ by substituting this in Eq. (9.3.17):

$$\int_{a}^{a+\rho} \frac{db}{\sqrt{(a+\rho)^2 - b^2}} = \cos^{-1}\left(\frac{a}{a+\rho}\right) \tag{9.3.18}$$

If we substitute Eq. (9.3.18) into Eq. (9.3.16) then we have:

$$K_I = \frac{2\sigma_p \sqrt{\pi(a+\rho)}}{\pi} \cos^{-1}\left(\frac{a}{a+\rho}\right) \tag{9.3.19}$$

Equation (9.3.19) is very important in the next section and by that we can explain Dugdale method for EPFM.

9.4 ANALYSIS OF EPFM AND CTOD USING DUGDALE METHOD

Dugdale method is similar to the one proposed by Irwin, but it is applicable to plane stress case only and can be explained by superposition principle.

Since in plane stress state, plastic region is much bigger and the yield stress σ_p can occupy a region ρ in each side of the crack, it is fare to assume that the remote stress σ which is applied to all the remote border (see Fig. 9.10) can stay in static equilibrium with σ_p in a bound region $a < b < a + \rho$.

This means if we apply a compressive stress to close the hypothetical crack with half length $a + \rho$, the resulted configuration has zero SIF since the net force is zero, this is described via superposition in Fig. 9.11.

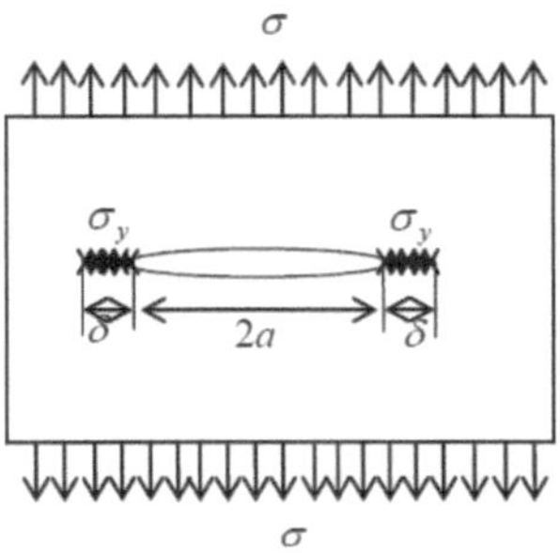

FIGURE 9.10 Plastic region and body.

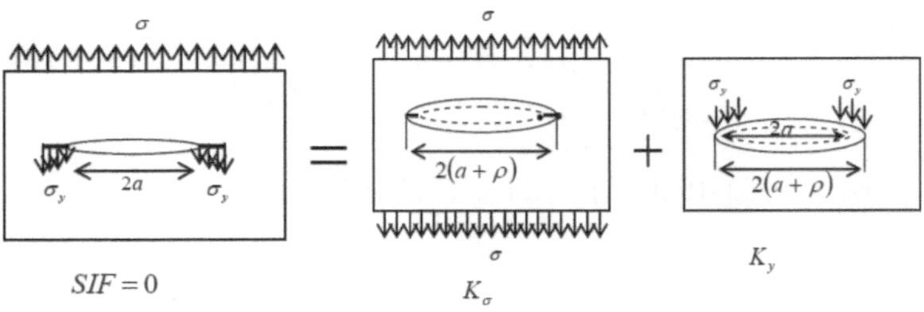

FIGURE 9.11 Superposition of remote stress and yield stress.

The compressive stress σ_p in the region $a < b < a + \rho$ wants to close the hypothetical region of the crack and it produces a negative SIF called K_Y and we studied its absolute value in previous section. Meanwhile the remote stress σ can produce a positive K_σ which is matching with crack length $2(a+\rho)$, the algebraic sum will be zero, that is,

$$K_Y + K_\sigma = 0 \tag{9.4.1}$$

The K_σ is:

$$K_\sigma = \sigma\sqrt{\pi(a+\rho)} \tag{9.4.2}$$

Subject to satisfaction of Eq. (9.4.1) we can use ELFM and analyze the problems in EPFM, K_Y is the same as Eq. (9.3.19), only we need to change to the minus sign, substituting Eq. (9.3.19) with minus sign and Eq. (9.4.2) into Eq. (9.4.1) we have:

$$\sigma\sqrt{\pi(a+\rho)} - \frac{2\sigma_p\sqrt{\pi(a+\rho)}}{\pi}\cos^{-1}\left(\frac{a}{a+\rho}\right) = 0 \tag{9.4.3}$$

If we divide Eq. (9.4.3) to $\sqrt{\pi(a+\rho)}$ then we have:

$\sigma - \frac{2\sigma_p}{\pi}\cos^{-1}\left(\frac{a}{a+\rho}\right) = 0 \Rightarrow \cos^{-1}\left(\frac{a}{a+\rho}\right) = \frac{\pi\sigma}{2\sigma_p}$ which finally results:

$$\frac{a}{a+\rho} = \cos\left(\frac{\pi\sigma}{2\sigma_p}\right) \tag{9.4.4}$$

In Eq. (9.4.4) if $\sigma << \sigma_p$ or $\frac{\sigma}{\sigma_p} << 1$ we can write:

$\cos\left(\dfrac{\pi\sigma}{2\sigma_p}\right)=1-2\sin^2\left(\dfrac{\pi\sigma}{4\sigma_p}\right)\cong 1-\dfrac{\pi^2\sigma^2}{8\sigma_p^2}$ and then Eq. (9.4.4) will change to:

$\dfrac{a}{a+\rho}=1-\dfrac{\rho}{a+\rho}\cong 1-\dfrac{\rho}{a}\cong 1-\dfrac{\pi^2\sigma^2}{8\sigma_p^2}$ which results an approximate formula for ρ:

$$\rho \cong \frac{\pi^2 a}{8}\left(\frac{\sigma}{\sigma_p}\right)^2 \tag{9.4.5}$$

If we use ELFM definition $K_I=\sigma\sqrt{\pi a}$ and substitute in (9.4.5) we have another form:

$$\rho \cong \frac{\pi}{8}\left(\frac{K_I}{\sigma_p}\right)^2 \tag{9.4.6}$$

If we compare Eq. (9.4.6) with Eq. (9.2.7), which gives Irwin's calculation for r_p, then we find out that $\rho > r_p$, now the question is how can we calculate δ or CTOD in this approach? For this we need to go to Chapter 7, and use Eq. (7.2.14) again which is:

$$K_I=\sqrt{HG}=\sqrt{EG} \text{ for plane stress} \tag{7.2.14}$$

Equation (7.2.14) is applicable to any case including K_Y and we can write:

$$K_Y^2=E\,G_Y \tag{9.4.7}$$

In Eq. (9.4.7) the G_Y is the strain energy release rate that is studied thoroughly in Chapter 7 and we rewrite again.

$$G_Y=\frac{\partial U_Y}{\partial(a+\rho)} \tag{9.4.8}$$

The strain energy can be expressed in terms of σ_p and the crack opening $v(a,b)$, a and b are explained in Section 9.3,

$$U_Y=\frac{1}{2}\int \sigma_p v_Y(b,a)\,db \tag{9.4.9}$$

By substituting Eq. (9.4.9) into Eq. (9.4.8) we have:

$$G_Y = \frac{\partial U_Y}{\partial(a+\rho)} = \frac{1}{2}\int \sigma_p \frac{\partial v_Y(b,a)}{\partial(a+\rho)}\, db \tag{9.4.10}$$

From Eq. (9.4.7) we have $G_Y = \frac{K_Y^2}{E}$ and if we substitute in Eq. (9.4.10) we have:

$$\frac{1}{2}\int \sigma_p \frac{\partial v_Y(b,a)}{\partial(a+\rho)}\, db = \frac{K_Y^2}{E} \tag{9.4.11}$$

By further manipulation (9.4.11) can change to:

$$K_Y = \int \sigma_p \frac{E}{2K_Y} \frac{\partial v_Y(b,a)}{\partial(a+\rho)}\, db \tag{9.4.12}$$

From Eq. (9.4.12), we define a Kernel function $m(a,b)$ like this:

$$m(a,b) = \frac{E}{2K_Y} \frac{\partial v_Y(b,a)}{\partial(a+\rho)} \tag{9.4.13}$$

Then Eq. (9.4.12) can be written in an abbreviated form:

$$K_Y = \int \sigma_p\, m(a,b)\, db \tag{9.4.14}$$

If we compare Eq. (9.4.14) with Eq. (9.3.16) we can find the Kernel function, that is,

$$m(b,a) = \sqrt{\frac{a+\rho}{\pi}} \frac{1}{\sqrt{(a+\rho)^2 - b^2}} \tag{9.4.15}$$

Equation (9.4.15) is necessary for the calculation of v_Y, we also need to calculate v_σ which is the result of remote stress σ and v_Y is the result of plastic stress σ_p. Determination of v_σ is straightforward from Chapter 7, Eq. (7.1.12) we write it again.

$$v(x) = \left(\frac{\kappa+1}{4\mu}\right)\sigma\sqrt{a^2 - x^2} \tag{7.1.12}$$

The parameters κ and μ are defined in chapter one by Eq. (6.4.26) written here again:

$$\mu=\frac{E}{2(1+\upsilon)} \qquad \kappa=\frac{3-\upsilon}{1+\upsilon} \tag{6.4.26}$$

Then we $\kappa+1=\frac{3-\upsilon}{1+\upsilon}+1=\frac{4}{1+\upsilon}$ for plane stress case and in Eq. (7.1.12) we need to replace $a \Leftrightarrow a+\rho$ and also $x \Leftrightarrow b$, then we have:

$$v_\sigma=\frac{2\sigma}{E}\sqrt{(\rho+a)^2-b^2} \tag{9.4.16}$$

If we examine Eq. (9.4.4) we have:

$$\frac{a}{a+\rho}=\cos\left(\frac{\pi\,\sigma}{2\sigma_p}\right)\Rightarrow\frac{\pi\,\sigma}{2\sigma_p}$$
$$=\cos^{-1}\left(\frac{a}{a+\rho}\right)\Rightarrow\sigma=\frac{2\sigma_p}{\pi}\cos^{-1}\left(\frac{a}{a+\rho}\right) \tag{9.4.17}$$

If we substitute Eq. (9.4.17) into Eq. (9.4.16) we have:

$$v_\sigma=\frac{4\sigma_p}{\pi\,E}\cos^{-1}\left(\frac{a}{a+\rho}\right)\sqrt{(\rho+a)^2-b^2} \tag{9.4.18}$$

If we compare Eq. (9.4.15) and Eq. (9.4.13) then we have:

$\frac{E}{2K_Y}\frac{\partial v_Y(b,a)}{\partial(a+\rho)}=\sqrt{\frac{a+\rho}{\pi}}\left((a+\rho)^2-b^2\right)^{-1/2}$ therefore, we can write:

$$\frac{\partial v_Y(b,a)}{\partial(a+\rho)}=\frac{2K_Y}{E}\sqrt{\frac{a+\rho}{\pi}}\left((a+\rho)^2-b^2\right)^{-1/2} \tag{9.4.19}$$

If we substitute Eq. (9.3.19) with negative sign into Eq. (9.4.19) then we have:

$$\frac{\partial v_Y(b,a)}{\partial(a+\rho)} = \frac{2}{E}\frac{-2\sigma_p\sqrt{\pi(a+\rho)}}{\pi}\cos^{-1}\left(\frac{a}{a+\rho}\right)\sqrt{\frac{a+\rho}{\pi}}\left((a+\rho)^2-b^2\right)^{-1/2}$$

simplified to:

$$\frac{\partial v_Y(b,a)}{\partial(a+\rho)} = \frac{-4\sigma_p}{\pi E}\cos^{-1}\left(\frac{a}{a+\rho}\right)\frac{a+\rho}{\sqrt{(a+\rho)^2-b^2}} \tag{9.4.20}$$

For determination of v_Y integration versus $(a+\rho)$ in Eq. (9.4.20) is required, that is,

$$v_Y = \frac{-4\sigma_p}{\pi E}\int\frac{(a+\rho)\cos^{-1}\left(\frac{a}{a+\rho}\right)}{\sqrt{(a+\rho)^2-b^2}}d(a+\rho) \tag{9.4.21}$$

Exercise 9.4.1: *Evaluate the above integral by using by part integration*

Solution: assume $\frac{(a+\rho)d(a+\rho)}{\sqrt{(a+\rho)^2-b^2}} = dA \Rightarrow A = \sqrt{(a+\rho)^2-b^2}$ and then $B = \cos^{-1}\left(\frac{a}{a+\rho}\right)$ results $dB = \frac{a\,d(a+\rho)}{(a+\rho)\sqrt{(a+\rho)^2-a^2}}$ the by part rule is $\int B\,dA = B\,A - \int A\,dB$ therefore Eq. (9.4.21) is:

$$\int\frac{(a+\rho)\cos^{-1}\left(\frac{a}{a+\rho}\right)}{\sqrt{(a+\rho)^2-b^2}}d(a+\rho) = \sqrt{(a+\rho)^2-b^2}\cos^{-1}\left(\frac{a}{a+\rho}\right) - \int\frac{\sqrt{(a+\rho)^2-b^2}}{\sqrt{(a+\rho)^2-a^2}}\left(\frac{a}{a+\rho}\right)d(a+\rho)$$

and this yields to:

$$v_Y = \frac{-4\sigma_p}{\pi E}\left[\sqrt{(a+\rho)^2-b^2}\cos^{-1}\left(\frac{a}{a+\rho}\right) - \int\frac{\sqrt{(a+\rho)^2-b^2}}{\sqrt{(a+\rho)^2-a^2}}\left(\frac{a}{a+\rho}\right)d(a+\rho)\right]$$

(9.4.22)

Dugdale method is based on this statement that, in order to use ELFM to analyze EPFM, the crack opening should be calculated from the following expression (superposition):

$$v = v_Y + v_\sigma \tag{9.4.23}$$

If we substitute Eqs. (9.4.18) and (9.4.22) into Eq. (9.4.23) then we have:

$$v = \frac{4\sigma_p}{\pi E}\left[\int \frac{\sqrt{(a+\rho)^2 - b^2}}{\sqrt{(a+\rho)^2 - a^2}}\left(\frac{a}{a+\rho}\right) d(a+\rho)\right] \tag{9.4.24}$$

In order to calculate v which finally leads to CTOD in Dugdale approach we need evaluate the integral in Eq. (9.4.24). The author used symbolic math toolbox in MATLAB and it was not successful after 2 hours CPU time.

Exercise 9.4.2: *Determine the integral in* Eq. (24) *by changing the variable to* $t = \frac{\sqrt{(a+\rho)^2 - b^2}}{\sqrt{(a+\rho)^2 - a^2}}$

Solution: Then we have $t^2 = \frac{(a+\rho)^2 - b^2}{(a+\rho)^2 - a^2}$ which results $(a+\rho)^2 (t^2 - 1) = t^2 a^2 - b^2$ and then $a + \rho = \left(\frac{t^2 a^2 - b^2}{t^2 - 1}\right)^{1/2}$, then we can express $d(a+\rho)$ in terms of dt to be able to integrate $d(a+\rho) = \frac{1}{2}\left(\frac{2a^2 t(t^2-1) - 2t(t^2a^2 - b^2)}{(t^2-1)^2}\right)\left(\frac{t^2 a^2 - b^2}{t^2 - 1}\right)^{-1/2} dt$ and after simplification $d(a+\rho) = \frac{-(a^2 - b^2)t}{(t^2-1)^2}\left(\frac{t^2 a^2 - b^2}{t^2 - 1}\right)^{-1/2} dt$, also $\frac{a}{a+\rho} = a\left(\frac{t^2 a^2 - b^2}{t^2 - 1}\right)^{-1/2}$ if we substitute all in Eq. (9.4.24) we have:

$$v = \frac{4\sigma_p}{\pi E}\left[\int t\, a\left(\frac{t^2a^2-b^2}{t^2-1}\right)^{-1/2}\frac{-\left(a^2-b^2\right)t}{\left(t^2-1\right)^2}\left(\frac{t^2a^2-b^2}{t^2-1}\right)^{-1/2}dt\right]$$ which can be simplified to:

$$v = \frac{4\sigma_p a}{\pi E}\left[\int\frac{-\left(a^2-b^2\right)t^2}{\left(t^2-1\right)\left(t^2a^2-b^2\right)}dt\right] \tag{9.4.25}$$

The integral in Eq. (9.4.25) is versus t, the author tried again to find the above integral by MATLAB and the result cannot be simplified.

Exercise 9.4.3: *Determine the integral in* Eq. (9.4.25) *by partial fraction method*

Solution: $\frac{-\left(a^2-b^2\right)t^2}{\left(t^2-1\right)\left(t^2a^2-b^2\right)} = \frac{\mathrm{A}}{t-1}+\frac{\mathrm{B}}{t+1}+\frac{\mathrm{M}}{ta-b}+\frac{\mathrm{N}}{ta+b}$, it is necessary to find $\mathrm{A,B,M,N}$ to do this we use the equivalence of the numerator, that is,

$$\frac{-\left(a^2-b^2\right)t^2}{\left(t^2-1\right)\left(t^2a^2-b^2\right)} = \frac{\left(t^2a^2-b^2\right)(t+1)\mathrm{A}+\left(t^2a^2-b^2\right)(t-1)\mathrm{B}+(ta+b)\left(t^2-1\right)\mathrm{M}+(ta-b)\left(t^2-1\right)\mathrm{N}}{\left(t^2-1\right)\left(t^2a^2-b^2\right)}$$

t^3 coefficient	$(\mathrm{A}+\mathrm{B})a^2+(\mathrm{M}+\mathrm{N})a=0$	I
t coefficient	$(\mathrm{A}+\mathrm{B})b^2-(\mathrm{M}+\mathrm{N})a=0$	II
t^2 coefficient	$(\mathrm{A}-\mathrm{B})a^2+(\mathrm{M}-\mathrm{N})b=-\left(a^2-b^2\right)$	III
t^0 coefficient	$(\mathrm{A}-\mathrm{B})b^2-(\mathrm{M}-\mathrm{N})b=0$	IV

III+IV results $(\mathrm{A}-\mathrm{B})\left(a^2-b^2\right)=-\left(a^2-b^2\right)\Rightarrow \mathrm{A}-\mathrm{B}=-1$ and I+II results $(\mathrm{A}+\mathrm{B})\left(a^2-b^2\right)=0\Rightarrow \mathrm{A}+\mathrm{B}=0$, then $\mathrm{A}={}^{-1}\!/_2$ and $\mathrm{B}={}^{1}\!/_2$, then we substitute and find out that $\mathrm{M}+\mathrm{N}=0$ and from III we have $\mathrm{M}-\mathrm{N}=b$, therefore $\mathrm{M}={}^{b}\!/_2$ and $\mathrm{N}={}^{-b}\!/_2$ by substituting $\mathrm{A,B,M,N}$ in Eq. (9.4.25) we can write:

$v = \frac{4\sigma_p a}{\pi E} \int \left[\frac{-dt}{2(t-1)} + \frac{dt}{2(t+1)} + \frac{b\,dt}{2(ta-b)} - \frac{b\,dt}{2(ta+b)} \right]$ and after performing the integration:

$$v = \frac{4\sigma_p a}{\pi E}\left(-\frac{1}{2}\ln(t-1) + \frac{1}{2}\ln(t+1) + \frac{b}{2a}\ln(at-b) - \frac{b}{2a}\ln(at+b) \right)$$

simplified form is:

$$v = \frac{4\sigma_p a}{\pi E}\left(\frac{1}{2}\ln\left(\frac{t+1}{t-1}\right) - \frac{b}{2a}\ln\left(\frac{t+b/a}{t-b/a}\right)\right) \quad (9.4.26)$$

Equation (9.4.26) is very important it enables the crack opening v to be calculated in any b, particularly if we want to calculate CTOD by Dugdale method we need to set $b \to a$ in Eq. (9.4.26)

$$\delta = \lim_{b\to a} 2v = \frac{8\sigma_p a}{\pi E}\left(\frac{1}{2}\ln\left(\frac{t+1}{t-1}\right) - \frac{b}{2a}\ln\left(\frac{t+b/a}{t-b/a}\right)\right) \quad (9.4.27)$$

If we set $b \to a$ this means that $t \to 1$ the above expression for δ, becomes ambiguous and to change the ambiguity we need to use l'Hopital's rule.

Exercise 9.4.4: *Remove the ambiguity from* Eq. (9.4.27) *by using l'Hopital's rule*

Solution: We express inside of the bracket in Eq. (9.4.27) into;

$$\lim_{\substack{b\to a\\ t\to 1}} \left(\frac{1}{2}\ln\left(\frac{t+1}{t-1}\right) - \frac{b}{2a}\ln\left(\frac{t+b/a}{t-b/a}\right) \right) = \lim_{\substack{b\to a\\ t\to 1}} \left(\frac{1}{2}\ln(t+1) - \frac{1}{2}\ln(t-1) - \frac{b}{2a}\ln\left(t+b/a\right) + \frac{b}{2a}\ln\left(t-b/a\right) \right)$$

$$\lim_{\substack{b\to a\\ t\to 1}}\left(\begin{array}{l}\frac{1}{2}\ln\left(\frac{t+1}{t-1}\right)-\\ \frac{b}{2a}\ln\left(\frac{t+b/a}{t-b/a}\right)\end{array}\right)=\lim_{\substack{b\to a\\ t\to 1}}\left(\begin{array}{l}\frac{1}{2}\ln(2)-\frac{1}{2}\ln(t-1)-\\ \frac{1}{2}\ln(2)+\frac{b}{2a}\ln\left(t-b/a\right)\end{array}\right)\text{ or}$$

$$\lim_{\substack{b\to a\\ t\to 1}}\left(\begin{array}{l}\frac{1}{2}\ln\left(\frac{t+1}{t-1}\right)-\\ \frac{b}{2a}\ln\left(\frac{t+b/a}{t-b/a}\right)\end{array}\right)=\lim_{\substack{b\to a\\ t\to 1}}\left(\begin{array}{l}-\frac{1}{2}\ln(t-1)+\frac{1}{2}\\ \ln\left(t-b/a\right)\end{array}\right)=\lim_{\substack{b\to a\\ t\to 1}}\left(\begin{array}{l}\frac{1}{2}\ln\\ \left(\frac{t-b/a}{t-1}\right)\end{array}\right)$$

Now we substitute instead of t in the above then we can remove ambiguity:

$$\lim_{\substack{b\to a\\ t\to 1}}\left(\frac{1}{2}\ln\left(\frac{t-b/a}{t-1}\right)\right)=\lim_{b\to a}\left(\frac{1}{2}\ln\left(\frac{\frac{\sqrt{(a+\rho)^2-b^2}}{\sqrt{(a+\rho)^2-a^2}}-b/a}{\frac{\sqrt{(a+\rho)^2-b^2}}{\sqrt{(a+\rho)^2-a^2}}-1}\right)\right) \tag{9.4.28}$$

Obviously Eq. (9.4.28) is still should be simplified into this:

$$\lim_{\substack{b\to a\\ t\to 1}}\left(\frac{1}{2}\ln\left(\frac{t-b/a}{t-1}\right)\right)=\lim_{b\to a}\left(\frac{1}{2}\ln\left(\frac{a\sqrt{(a+\rho)^2-b^2}-b\sqrt{\rho^2+2a\rho}}{\left(\sqrt{(a+\rho)^2-b^2}-\sqrt{\rho^2+2a\rho}\right)a}\right)\right),$$

then the problem reduces to removing ambiguity from inner bracket, that is, $\lim_{b\to a}\frac{a\sqrt{(a+\rho)^2-b^2}-b\sqrt{\rho^2+2a\rho}}{\left(\sqrt{(a+\rho)^2-b^2}-\sqrt{\rho^2+2a\rho}\right)a}$, for using Hoptial rule we need to differentiate versus b.

$$\lim_{b\to a}\frac{a\sqrt{(a+\rho)^2-b^2}-b\sqrt{\rho^2+2a\rho}}{\left(\sqrt{(a+\rho)^2-b^2}-\sqrt{\rho^2+2a\rho}\right)a}=\lim_{b\to a}\frac{\frac{-ab}{\sqrt{(a+\rho)^2-b^2}}-\sqrt{\rho^2+2a\rho}}{\frac{-ab}{\sqrt{(a+\rho)^2-b^2}}}$$

$$\lim_{b\to a}\frac{\dfrac{-ab}{\sqrt{(a+\rho)^2-b^2}}-\sqrt{\rho^2+2a\rho}}{\dfrac{-ab}{\sqrt{(a+\rho)^2-b^2}}}=$$

$$\lim_{b\to a}\frac{-ab-\sqrt{\rho^2+2a\rho}\sqrt{(a+\rho)^2-b^2}}{-ab}=\left(\frac{a+\rho}{a}\right)^2$$ If we substitute the above into Eq. (9.4.28) the 1/2 inside the bracket will be removed and Eq. (9.4.27) is:

$$\delta=\frac{8\sigma_p a}{\pi E}\left(\ln\left(\frac{a+\rho}{a}\right)\right) \tag{9.4.29}$$

By inverting equation (9.4.4) we have $\frac{a+\rho}{a}=\sec\left(\frac{\pi\sigma}{2\sigma_p}\right)$ and if we substitute in (9.4.29).

$$\delta=\frac{8\sigma_p a}{\pi E}\left(\ln\left(\sec\left(\frac{\pi\sigma}{2\sigma_p}\right)\right)\right) \tag{9.4.30}$$

The above equation gives the CTOD in Dugdale method and has an interesting features particularly when $\frac{\sigma}{\sigma_p}<<1$ or $\sigma<<\sigma_p$, the term inside bracket can be simplified, that is,

$$\ln\left(\sec\left(\frac{\pi\sigma}{2\sigma_p}\right)\right)\cong\frac{\pi^2\sigma^2}{8\sigma_p^2} \tag{9.4.31}$$

If we substitute Eq. (9.4.31) into Eq. (9.4.30) then we have:

$$\delta\cong\frac{\pi a\sigma^2}{E\sigma_p} \tag{9.4.32}$$

However, we know that $K_I=\sigma\sqrt{\pi a}$ and this change Eq. (9.4.32) into $\delta\cong\frac{K_I^2}{E\sigma_p}$, also in Chapter 2, we saw that $G=\frac{K_I^2}{E}$ and this simplifies even further so that $\delta \cong \dfrac{G}{\sigma_p}$. This leads to a very important equation in EPFM, saying that critical CTOD δ_C is simply related to critical energy release rate G_C via following simple relationship:

$$G_c \cong \sigma_p \delta_c \tag{9.4.33}$$

The CTOD first introduced by Eq. (9.4.33) in UK, and Irwin was not aware of that. This equation saying that in EPFM, the critical energy release rate G_c is proportional to δ_c via the yield stress σ_p, in the simplest case. In general case we can conclude that: $G_c = \alpha \sigma_p \delta_c$, and in Section 9.2 we saw in Eq. (9.2.17) that $\alpha = \pi / 4$.

KEYWORDS

- **crack tip opening displacement**
- **elastic-plastic fracture mechanic**
- **Hook's law**
- **l'Hopital's Rule**
- **linear-elastic fracture mechanic**
- **Mohr circle**
- **nonlinear fracture mechanics**
- **stress distribution**
- **Von Mises law**

CHAPTER 10

STRESS INTENSITY FACTOR FOR THROUGH THICKNESS FLAWS AND J INTEGRAL

CONTENTS

10.1 WESTERGAARD FORMULAS FOR DISPLACEMENTS AROUND CRACK TIP

In this section, the aim is studying about the crack opening v, around the crack tip. We need to use the results of this section to analyse three-dimensional cracks. It necessary to rewrite Eq. (6.4.27), which is:

$$2\mu(u+iv)=\kappa\phi(z)-z\overline{\phi'(z)}-\overline{\psi'(z)} \tag{6.4.27}$$

Also in Chapter 7, when we studied Westergaard functions, we said that $\psi'(z)$ and $\varphi(z)$ are related via Eq. (7.3.5) as follows:

$$\psi'(z)=\phi(z)-z\phi'(z)+Az \tag{7.3.5}$$

Moreover, we rewrite the definition of Westergaard function as well.

$$Z_I(z)=2\phi'(z) \tag{7.3.8}$$

$$\tilde{Z}_I(z) = \int Z_I(z)\,dz = 2\phi(z) \tag{7.3.10}$$

According to Eq. (7.3.5) we can find out $\overline{\psi'(z)}$ as follows:

$$\overline{\psi'(z)} = \overline{\phi(z)} - \overline{z\phi'(z)} + A\bar{z} \tag{10.1.1}$$

In this section we want to determine v only, then it is necessary to evaluate all the imaginary terms in Eq. (10.1.1), that is,

$$\Im\left(-\overline{\phi(z)}\right) = \Im\left(\Re(-\phi(z)) - i\Im(-\phi(z))\right) = \Im(\phi(z)) \tag{10.1.2}$$

$$\begin{aligned}\Im\left(\overline{z\phi'(z)}\right) &= \Im\left(\overline{(x+iy)\left(\Re(\phi'(z)) + i\Im(\phi'(z))\right)}\right)\\ &= -x\,\Im(\phi'(z)) - y\,\Re(\phi'(z))\end{aligned} \tag{10.1.3}$$

$$\Im\left(A\bar{z}\right) = -Ay \tag{10.1.4}$$

If we substitute Eq. (10.1.1) into Eq. (6.4.27) we have:

$$2\mu(u+iv) = \kappa\phi(z) - z\overline{\phi'(z)} - \overline{\phi(z)} + \overline{z\phi'(z)} - A\bar{z} \tag{10.1.5}$$

By looking at Eq. (10.1.5) it is obvious that imaginary parts of $\kappa\phi(z)$ and also $-\overline{z\phi'(z)}$ are necessary for determination of the crack opening or v and they are:

$$\Im(\kappa\phi(z)) = \kappa\left(\Re(\phi(z)) + i\Im(\phi(z))\right) = \kappa\,\Im(\phi(z)) \tag{10.1.6}$$

$$\begin{aligned}\Im\left(-\overline{z\phi'(z)}\right) &= \Im\left((-x-iy)\left(\Re(\phi'(z)) + i\Im(\phi'(z))\right)\right)\\ &= x\,\Im(\phi'(z)) - y\,\Re(\phi'(z))\end{aligned} \tag{10.1.7}$$

By substituting Eqs. (10.1.2), (10.1.3), (10.1.4), (10.1.6) and (10.1.7) into Eq. (10.1.5) we have:

$2\mu v = \kappa \Im(\phi(z)) + x\Im(\phi'(z)) - y\Re(\phi'(z)) + \Im(\phi(z)) - x\Im(\phi'(z)) - y$ $\Re(\phi'(z)) + Ay$, that can be simplified to:

$$2\mu v = (\kappa+1)\Im(\phi(z)) - 2y\Re(\phi'(z)) + Ay \qquad (10.1.8)$$

If we substitute Eqs. (7.3.8) and (7.3.10) into Eq. (10.1.8) we have a formula in terms of Westergaard functions, i.e.

$$\mu v = \left(\frac{\kappa+1}{4}\right)\Im(\tilde{Z}_I(z)) - \frac{y}{2}\Re(Z_I(z)) + \frac{A}{2}y \qquad (10.1.9)$$

For $\tilde{Z}_I(z)$ in chapter two we use Eq. (7.1.6) given for $\varphi(z)$ and find out that:

$$\tilde{Z}_I(z) = \frac{\sigma}{2}\left(2\sqrt{z^2-a^2} - z\right) \qquad (10.1.10)$$

For plane strain case $\kappa = 3 - 4\upsilon$, also $\mu = G$ or shear modulus (not energy release rate), then Eq. (10.1.9) will change to:

$$v = \left(\frac{1-\upsilon}{G}\right)\Im(\tilde{Z}_I(z)) - \frac{y}{2G}\Re(Z_I(z)) + \frac{A}{2G}y \qquad (10.1.11)$$

Now for a crack we write $Z_I(z)$ by using Eq. (7.3.17) in chapter two, i.e.

$$Z_I(z) = 2\phi'(z) = \frac{\sigma}{2}\left(\frac{2z}{\sqrt{z^2-a^2}} - 1\right) \qquad (7.3.17)$$

From (7.3.24) we had $A = \frac{\sigma}{2}$, substituting this Eq. (10.1.10) and also Eq. (7.3.17) in Eq. (10.1.11) results this:

$$v = \left(\frac{1-\upsilon}{G}\right)\Im\left(\frac{\sigma}{2}\left(2\sqrt{z^2-a^2} - z\right)\right) - \frac{y}{2G}\Re\left(\frac{\sigma}{2}\left(\frac{2z}{\sqrt{z^2-a^2}} - 1\right)\right) + \frac{y}{2G}\frac{\sigma}{2}$$

which can be simplified to:

$$v = \frac{\sigma}{2G}\left((1-\upsilon)\Im\left(2\sqrt{z^2-a^2} - z\right) - \frac{y}{2}\Re\left(\frac{2z}{\sqrt{z^2-a^2}}\right)\right) \qquad (10.1.12)$$

If we want to use Eq. (10.1.12), to determine the crack opening in the vicinity of the crack tip, similar to chapter 8, we should consider $z-a=re^{i\theta}$, also $z+a\cong 2a$ and $z\cong a$ assuming that in vicinity of the crack $r<<a$. Then we can approximate the individual terms in Eq. (10.1.12) in vicinity of the crack tip.

$\Im\left(2\sqrt{z^2-a^2}-z\right)\cong\Im\left(2\sqrt{z-a}\sqrt{z+a}-a\right)=\Im\left(2\sqrt{2a}\sqrt{re^{i\theta}}\right)=$

$2\sqrt{2a}\sqrt{r}\,\sin\frac{\theta}{2}$, therefore, we can write:

$$\Im\left(2\sqrt{z^2-a^2}-z\right)\cong 2\sqrt{2a}\sqrt{r}\,\sin\frac{\theta}{2} \tag{10.1.13}$$

Also we have:

$$\frac{y}{2}=\frac{r\sin\theta}{2}=r\sin\frac{\theta}{2}\cos\frac{\theta}{2} \tag{10.1.14}$$

$\Re\left(\frac{2z}{\sqrt{z^2-a^2}}\right)\cong\Re\left(\frac{2a}{\sqrt{z-a}\sqrt{z+a}}\right)\cong\Re\left(\frac{2a}{\sqrt{re^{i\theta}}\sqrt{2a}}\right)$ and if we simplify this we have:

$$\Re\left(\frac{2z}{\sqrt{z^2-a^2}}\right)\cong\sqrt{\frac{2a}{r}}\cos\frac{\theta}{2} \tag{10.1.15}$$

If we substitute Eq. (10.1.13), (10.1.14) and (10.1.15) into Eq. (10.1.12) then we have:

$$v=\frac{\sigma}{2G}\left(2(1-\upsilon)\sqrt{2a}\sqrt{r}\,\sin\frac{\theta}{2}-r\sin\frac{\theta}{2}\cos\frac{\theta}{2}\sqrt{\frac{2a}{r}}\cos\frac{\theta}{2}\right) \tag{10.1.16}$$

The stress σ, can be expressed in terms of SIF, or $K_I=\sigma\sqrt{\pi a}$ if we substitute in Eq. (10.1.16) we have:

$$v=\frac{K_I}{G}\sqrt{\frac{r}{2\pi}}\sin\frac{\theta}{2}\left(2-2\upsilon-\cos^2\frac{\theta}{2}\right) \tag{10.1.17}$$

In Eq. (10.1.17) the normal displacement v in any point near crack tip can be calculated. To find the opening of the crack faces we need to set $\theta = \pi$, in Eq. (10.1.17) and this has been clarified in figure 1, here $\sin\frac{\pi}{2} = 1$ and then Eq. (10.1.17) becomes:

$$v = \frac{2K_I}{G}\sqrt{\frac{r}{2\pi}}(1-\upsilon) \tag{10.1.18}$$

Equation (10.1.18) is used by Irwin to study three-dimensional cracks, for isotropic materials he used the relationship between shear modulus and Young modulus, that is, $\frac{1}{G} = \frac{2(1+\upsilon)}{E}$, and substituted into (10.1.18) then the result is:

$$v = \frac{2(1-\upsilon^2)}{E}K_I\sqrt{\frac{2r}{\pi}} \tag{10.1.19}$$

Although Irwin was the person who defined SIF, sometimes he has used another definition, which is $K_I = \frac{K_I}{\sqrt{\pi}}$, and in this case the Eq. (10.1.19) will change. We can move to next section now to study three-dimensional cracks.

10.2 ANALYSIS OF THE THREE-DIMENSIONAL CRACKS

Cracks in reality are three-dimensional, and approximating them by two-dimensional slots is a simplification for the easy analysis. Mathematicians have studied three-dimensional cracks, before but they did not achieve practical solutions. Irwin is the first who studied three-dimensional cracks with elliptic cross section. He derived formulas, which are still used in fracture mechanics according to Fig. 10.2.

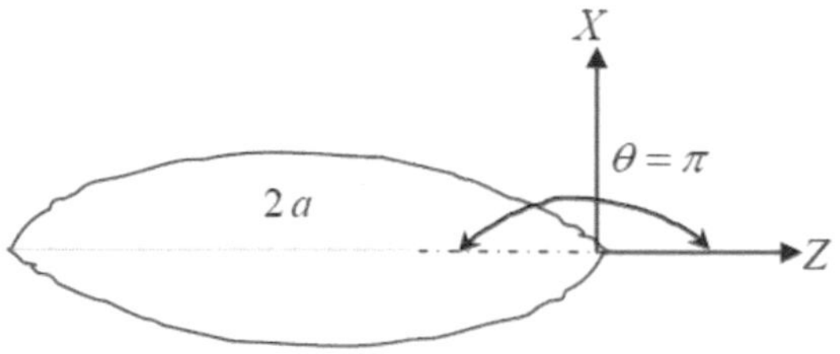

FIGURE 10.1 Crack face opening.

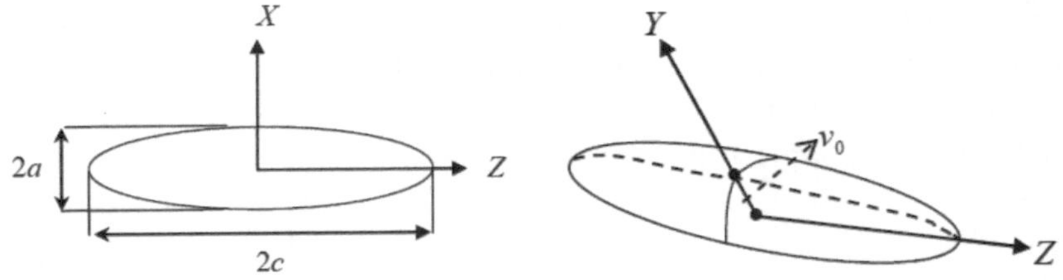

FIGURE 10.2 Elliptical crack.

Three-dimensional cracks with elliptical cross sections and major diameter $2c$ and minor diameter $2a$, under the loading will take an ellipsoidal shape like egg. The crack opening in the centre of ellipse is maximum given by $2v_0$ (v_0 in each face) and in the border of the crack, which is expressed by this ellipse:

$$\frac{x_1^2}{a^2}+\frac{z_1^2}{c^2}=1 \tag{10.2.1}$$

Irwin assumed z axis on major diameter and x on minor diameter and the y axis in direction of the crack opening v. The points x_1, z_1 are on the crack border, while x, z are internal points of the crack. Irwin assumed the opening function like this:

$$v^2=\left(1-\frac{x^2}{a^2}-\frac{z^2}{c^2}\right)v_0^2 \tag{10.2.2}$$

The mathematicians before Irwin did assume the same opening function in Eq. (10.2.2), but until 1962, no practical solution was available. He discussed that even if the method of solution is approximate but if the solution satisfies special cases then it can be acceptable.

According to Figs. 10.3 and 10.4, each point P on the ellipse can be measured relative to a big circle (radius c) and a small circle (radius b). If we draw a normal line into the z axis and name it PH, then PH crosses the big circle at point S, then connect the S to O (the ellipse centre) so that SO, forms an angle ϕ with the z axis then we have:

$$OH=z_1=c\cos\phi \tag{10.2.3}$$

$$PH=x_1=a\sin\phi=MK \tag{10.2.4}$$

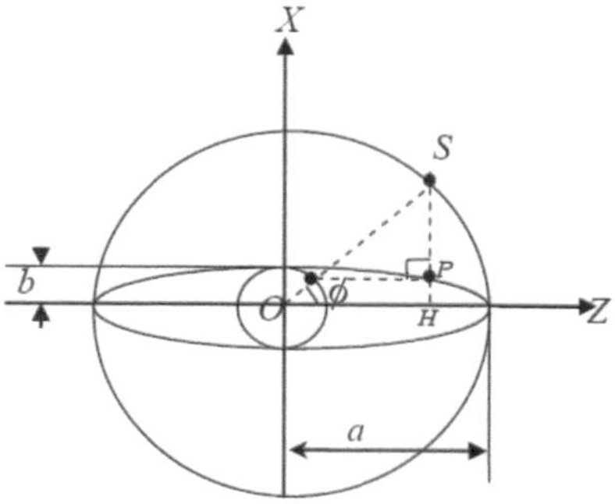

FIGURE 10.3 Big and small circle.

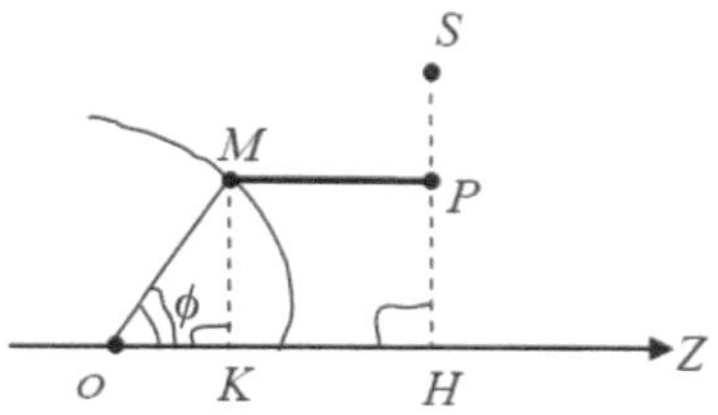

FIGURE 10.4 The points S, M, P, K and H.

If we draw a line from P parallel to z axis to intersect the small circle in point M, then from this point draw a normal line to z axis, named MK (Fig. 10.5). Then OM and PH intersect each other at point S' if OM makes an angle θ with the z axis then we have:

$\frac{OK}{OH}=\frac{a\cos\theta}{c\cos\theta}=\frac{OM}{OS'}=\frac{a}{OS'}\Rightarrow OS'=c$ and this confirms that the point S' is on the big circle and coincides with the point S, therefore it can be proved that Eqs. (10.2.3) and (10.2.4) are correct. Moreover, Eqs. (10.2.3) and (10.2.4) satisfies the equation of the crack border in Eq. (10.2.1). If we differentiate from Eqs. (10.2.3) and (10.2.4) versus ϕ then we have:

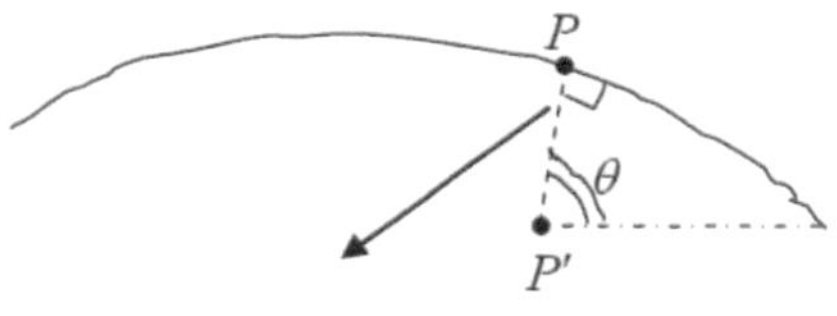

FIGURE 10.5 Internal and border points.

$$dz_1 = -c \sin\phi \, d\phi \qquad dx_1 = a \cos\phi \, d\phi \tag{10.2.5}$$

The infinitesimal crack length on the border is designate by *ds* and can be found by:

$$ds = \sqrt{dx_1^2 + dz_1^2} \Rightarrow ds = \sqrt{c^2 \sin^2\phi + a^2 \cos^2\phi} \, d\phi \tag{10.2.6}$$

Later we will use Eq. (10.2.6), the important work that Irwin did was the conversion of Eq. (10.2.22) in terms of the variable ϕ, which is not an easy task and the details of his derivation is not published anywhere.

The author believes that according to Fig. 10.5, for any point P on the elliptical border of the crack, a normal is drawn to the ellipse and then across this normal toward the centre, all the internal points $P'(x,z)$ can be defined versus θ, the angle of the normal with the z axis and also r, the distance $P'P$ across the normal line. Then we have:

$$z = z_1 - r\cos\theta \tag{10.2.7}$$

$$x = x_1 - r\sin\theta \tag{10.2.8}$$

We will show later in the section, that he assumed that r in Eqs. (10.2.7) and (10.2.8) is the same as r in Eq. (10.1.19) and he can use that formula for two-dimensional cracks here, in which the r represents all internal points of the three-dimensional crack. However, the complexity in Eqs. (10.2.7) and (10.2.8) is the angle θ that expresses the direction normal to the elliptical contours and is different from ϕ. Dividing x by a and z by c results:

$$\frac{x}{a} = \frac{x_1}{a} - \frac{r}{a}\sin\theta \tag{10.2.9}$$

$$\frac{z}{c} = \frac{z_1}{c} - \frac{r}{c}\cos\theta \tag{10.2.10}$$

Squaring both sides of Eqs. (10.2.9) and (10.2.10) and adding up together gives:

$$\frac{x^2}{a^2} + \frac{z^2}{c^2} = \frac{x_1^2}{a^2} + \frac{r^2}{a^2}\sin^2\theta - \frac{2x_1 r}{a^2}\sin\theta + \frac{z_1^2}{c^2} + \frac{r^2}{c^2}\cos^2\theta - \frac{2z_1 r}{c^2}\cos\theta$$

From Eqs. (10.2.3) and (10.2.4) we have $\frac{x_1}{a} = \sin\phi$ and $\frac{z_1}{c} = \cos\phi$ also we have $\frac{x_1^2}{a^2} + \frac{z_1^2}{c^2} = 1$ and substituting in the above expression results:

$$\frac{x^2}{a^2} + \frac{z^2}{c^2} = 1 + \frac{r^2}{a^2}\sin^2\theta - \frac{2r}{a}\sin\theta\sin\phi + \frac{r^2}{c^2}\cos^2\theta - \frac{2r}{c}\cos\theta\cos\phi \tag{10.2.11}$$

In Eq. (10.2.11) if $r << a$ then we can ignore the terms $\frac{r^2}{a^2}$ and $\frac{r^2}{c^2}$ when compared with 1, so $\frac{r^2}{a^2}\sin^2\theta \cong 0$ and also $\frac{r^2}{c^2}\cos^2\theta \cong 0$ then Eq. (10.2.11) can be approximated by:

$$1 - \frac{x^2}{a^2} - \frac{z^2}{c^2} \cong \frac{2r}{a}\sin\theta\sin\phi + \frac{2r}{c}\cos\theta\cos\phi \tag{10.2.12}$$

Exercise 10.2.1: Find an expression for θ direction normal to the elliptical contour

Solution: From Eq. (10.2.1) the expression for the elliptical border is:

$$x_1 = a\sqrt{1 - \frac{z_1^2}{c^2}} \tag{10.2.13}$$

Differentiation of Eq. (10.2.13) results:

$$\frac{dx_1}{dz_1} = \frac{-a z_1}{c^2\sqrt{1 - \frac{z_1^2}{c^2}}} \tag{10.2.14}$$

The equation in Eq. (10.2.14) expresses the tangent, and normal line makes the following angle:

$$\tan\theta = \frac{c^2\sqrt{1 - \frac{z_1^2}{c^2}}}{a z_1} \tag{10.2.15}$$

In Eq. (10.2.15) we can substitute $z_1 = c\cos\phi$ and then we have:

$$\tan\theta = \frac{c}{a}\tan\phi \tag{10.2.16}$$

$$\sin\theta = \frac{c\tan\phi}{\sqrt{a^2 + c^2\tan^2\phi}} \tag{10.2.17}$$

$$\cos\theta = \frac{a}{\sqrt{a^2 + c^2\tan^2\phi}} \tag{10.2.18}$$

By substituting Eqs. (10.2.17) and (10.2.18) into Eq. (10.2.12) we have:

$$\frac{2r}{ac}\left(\frac{c^2\tan\phi\sin\phi}{\sqrt{a^2 + c^2\tan^2\phi}} + \frac{a^2\cos\phi}{\sqrt{a^2 + c^2\tan^2\phi}}\right) = 1 - \frac{x^2}{a^2} - \frac{z^2}{c^2} \tag{10.2.19}$$

Using trigonometry we can simplify the (10.2.19) into:

$$\frac{2r}{ac}\left(\frac{c^2\sin^2\phi}{\sqrt{a^2\cos^2\phi + c^2\sin^2\phi}} + \frac{a^2\cos^2\phi}{\sqrt{a^2\cos^2\phi + c^2\sin^2\phi}}\right) = 1 - \frac{x^2}{a^2} - \frac{z^2}{c^2}$$

Which, can be simplified to:

$$\left(\sqrt{a^2\cos^2\phi + c^2\sin^2\phi}\right)\frac{2r}{ac} = 1 - \frac{x^2}{a^2} - \frac{z^2}{c^2} \tag{10.2.20}$$

Substituting Eq. (10.2.20) into Eq. (10.2.2) yields to:

$$v^2 = v_0^2\left(\sqrt{a^2\cos^2\phi + c^2\sin^2\phi}\right)\frac{2r}{ac} \tag{10.2.21}$$

r in Eq. (10.2.21) can be the same as the same as r in (Eq. 10.1.19) rewritten here:

$$v = \frac{2\left(1-\upsilon^2\right)}{E}K_I\sqrt{\frac{2r}{\pi}} \tag{10.2.22}$$

Squaring Eq. (10.2.22) gives $v^2 = \frac{4(1-\upsilon^2)^2}{E^2} K_I^2 \frac{2r}{\pi}$ and substituting into Eq. (10.2.21) results:

$$\frac{4(1-\upsilon^2)^2}{E^2} K_I^2 \frac{2r}{\pi} = v_0^2 \left(\sqrt{a^2 \cos^2 \phi + c^2 \sin^2 \phi}\right) \frac{2r}{ac}$$

Then SIF can be calculated from the above by:

$$K_I^2 = \frac{\pi}{4} \left(\frac{E}{1-\upsilon^2}\right)^2 \left(\frac{v_0^2}{ac}\right) \left(\sqrt{a^2 \cos^2 \phi + c^2 \sin^2 \phi}\right) \qquad (10.2.23)$$

Considering the fact that energy release rate in plane strain case can be calculated by $G_I = \frac{K_I^2}{H} = \frac{K_I^2}{\frac{E}{1-\upsilon^2}}$ and by substituting Eq. (10.2.23) into this we can write:

$$G_I = \frac{\pi}{4} \left(\frac{E}{1-\upsilon^2}\right) \left(\frac{v_0^2}{ac}\right) \left(\sqrt{a^2 \cos^2 \phi + c^2 \sin^2 \phi}\right) \qquad (10.2.24)$$

Equation (10.2.24) is very important and in fact Irwin used this equation for determination of the energy, which is required to open a three-dimensional elliptical crack as follows here.

The energy required for the crack opening by the amount of Δ in normal direction to its elliptic boundary (see Fig. 10.6) can be calculated by:

$$\delta U = 4 \int_0^{\pi/2} G_I \Delta \frac{ds}{d\phi} d\phi \qquad (10.2.25)$$

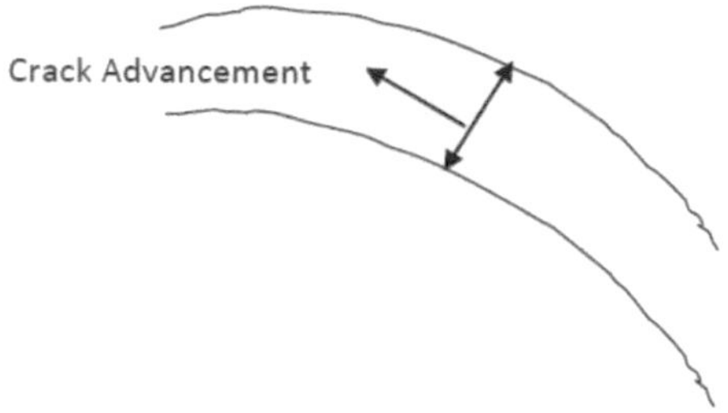

FIGURE 10.6 Elliptical crack opening.

The crack opening Δ moving outward the elliptical boundary and $\Delta\, ds$ is the infinitesimal area swept by the crack movement, that can be integrated over the ellipse periphery. It is necessary to define a function $\Delta(\phi)$ to be able to perform the integration in Eq. (10.2.25). It is obvious that in $\phi=0$, the crack opening is $\Delta(0)=c\,f$ in which $f \ll 1$ and very small. Similarly in $\phi=\pi/2$, the crack opening is $\Delta(\pi/2)=a\,f$ also the same $f \ll 1$. Considering these Irwin proposed the following function for $\Delta(\phi)$:

$$\Delta(\phi)=\frac{a\,c\,f}{\sqrt{a^2\cos^2\phi+c^2\sin^2\phi}} \tag{10.2.26}$$

From Eq. (10.2.26) it can be found that both $\Delta(0)=c\,f$ and also $\Delta(\pi/2)=a\,f$ and it is an appropriate function to be used. Now we can substitute Eqs. (10.2.26), (10.2.24) and (10.2.6) into Eq. (10.2.25) we have:

$$\delta U=4\int_0^{\pi/2}\frac{\pi}{4}\left(\frac{E}{1-\upsilon^2}\right)\left(\frac{v_0^2}{ac}\right)\left(\sqrt{a^2\cos^2\phi+c^2\sin^2\phi}\right)\frac{a\,c\,f}{\sqrt{a^2\cos^2\phi+c^2\sin^2\phi}}$$

$\sqrt{c^2\sin^2\phi+a^2\cos^2\phi}\;d\phi$ and can be simplified to $\delta U=\dfrac{\pi\,E\,v_0^2 f}{1-\upsilon^2}\displaystyle\int_0^{\pi/2}$ $\sqrt{c^2\sin^2\phi+a^2\cos^2\phi}\;d\phi$ or an alternative form:

$$\delta U=\frac{\pi\,E\,v_0^2 f\,c}{1-\upsilon^2}\int_0^{\pi/2}\sqrt{\sin^2\phi+\frac{a^2}{c^2}\cos^2\phi}\;d\phi=\frac{\pi\,E\,v_0^2 f\,c\,\Phi}{1-\upsilon^2} \tag{10.2.27}$$

In Eq. (10.2.27) Φ, is known as elliptic integral and obviously can be calculated by:

$$\Phi=\int_0^{\pi/2}\sqrt{\sin^2\phi+\frac{a^2}{c^2}\cos^2\phi}\;d\phi \tag{10.2.28}$$

Irwin calculated the energy release in Eq. (10.2.27) from another approach saying that δU is also depends on the $d\,V_c$ which is the volume change of the elliptical crack due to its opening. This can be measured by $f \ll 1$ if we assume that $dV_c = V_c f$. However, the crack volume can be found from $V_c = \underbrace{(\pi\, a c)}_{area}\underbrace{2v_0}_{hight}$ then we can write:

$$\delta U = 2\pi\, a c v_0 f \sigma = \underbrace{dV_c}_{volume\, change}\underbrace{\sigma}_{stress} \tag{10.2.29}$$

By equating δU from Eq. (10.2.27) and δU from Eq. (10.2.29) we have:

$2\pi\, a c v_0 f \sigma = \dfrac{\pi E v_0^2 f c \Phi}{1-\upsilon^2}$ and from this we can find an expression for v_0 which is the maximum crack opening and occurs at ellipse centre by:

$$v_0 = \frac{2a\left(1-\upsilon^2\right)\sigma}{E\Phi} \tag{10.2.30}$$

Now we substitute Eq. (10.2.30) into Eq. (10.2.24) then we have:

$G_I = \dfrac{\pi}{4}\left(\dfrac{E}{1-\upsilon^2}\right)\left(\dfrac{2a\left(1-\upsilon^2\right)\sigma}{E\Phi}\right)^2\left(\dfrac{1}{ac}\right)\left(\sqrt{a^2\cos^2\phi + c^2\sin^2\phi}\right)$ and can be simplified to:

$$G_I = \frac{\pi\left(1-\upsilon^2\right)\sigma^2}{E\Phi^2}\left(\frac{a}{c}\right)\left(\sqrt{a^2\cos^2\phi + c^2\sin^2\phi}\right) \tag{10.2.31}$$

Also for plane strain state we had $\left(\dfrac{E}{1-\upsilon^2}\right)G_I = K_I^2$ and by substituting Eq. (10.2.31) into this we have:

$$K_I^2 = \frac{\pi\, a\sigma^2}{c\Phi^2}\left(\sqrt{a^2\cos^2\phi + c^2\sin^2\phi}\right) \tag{10.2.32}$$

From Eq. (10.2.32) it is obvious that SIF for elliptical crack locally depends on the angle ϕ which for three-dimensional crack is expectable. The maximum SIF occurs at angle $\phi = \pi/2$ and substituting it in Eq. (10.2.32) gives:

$$K_{I\max}^2 = \frac{\pi\, a\sigma^2}{\Phi^2} \tag{10.2.33}$$

Although the procedure of calculating SIF by Eqs. (10.2.32) and (10.2.33), is an approximation but it satisfies the special cases. For example consider a circular crack in which $a = c$ that makes the elliptic integral in Eq. (10.2.28), $\Phi = \pi/2$ and by substituting this into Eq. (10.2.33) and taking the square root we have:

$$K_I = \frac{2}{\pi}\sigma\sqrt{\pi a} \tag{10.2.34}$$

The Eq. (10.2.34) was derived before Irwin for penny shaped (circular) cracks and can be approved by Eq. (10.2.32) as well. In other case if we consider two-dimensional cracks in which $a << c$ that makes the elliptic integral in Eq. (10.2.28), $\Phi = 1$ and by substituting this into Eq. (10.2.33) and taking the square root we have:

$$K_I = \sigma\sqrt{\pi a} \tag{10.2.35}$$

The Eq. (10.2.35) is the definition of SIF given by Irwin based on the Griffith theory of fracture and can be approved by Eq. (10.2.32) as well. For three-dimensional cracks Irwin always emphasized on $K_{I\max}$ given by Eq. (10.2.33) and names it K_I and he summarises his theory by theses formulas which are based on Eq. (10.2.33):

$$\underbrace{K_I^2\,\Phi^2 = \pi a \sigma^2}_{\text{embedded flaw}} \qquad \underbrace{K_I^2\,\Phi^2 = 1.21\pi a \sigma^2}_{\text{surface flaw}} \tag{10.2.36}$$

In Eq. (10.2.36) for the surface flaw there is a factor 1.21 that is inserted based on the theory for two-dimensional cracks, saying that "K_I^2 in surface cracks will be amplified by 1.21" this is proved by mathematicians before Irwin and will not be discussed here.

Moreover, he modifies his formulas to consider the effect of plastic deformation. To do this, in Eq. (10.2.36) he changes a to $a + r_y$ and considers the $r_y = \left(\frac{1}{4\pi\sqrt{2}}\right)\left(\frac{K_I}{\sigma_p}\right)^2$ (with no proof) for plane strain state. Therefore the modified Eq. (10.2.36) will be: $K_I^2\,\Phi^2 = 1.21\pi a \sigma^2 \left(a + \underbrace{\frac{K_I^2}{4\pi\sqrt{2}\sigma_p^2}}_{r_y} \right)$

which can be rearranged into: $K_I^2\,\Phi^2 - \dfrac{1.21}{4\sqrt{2}} K_I^2 \left(\dfrac{\sigma^2}{\sigma_p^2}\right) = 1.21\pi a\sigma^2$

which is the basis for two famous formulas for three-dimensional elliptical cracks as follows:

$$\underbrace{K_I^2 = \frac{1.21\pi a\sigma^2}{\Phi^2 - 0.213\left(\dfrac{\sigma}{\sigma_p}\right)^2}}_{surface\ flaw} \qquad \underbrace{K_I^2 = \frac{\pi a\sigma^2}{\Phi^2 - 0.213\left(\dfrac{\sigma}{\sigma_p}\right)^2}}_{embedded\ flaw} \tag{10.2.37}$$

10.3 *J* INTEGRAL AND THE RELATIONSHIP WITH G_I

As we saw in previous chapters, there are some important parameters in fracture mechanics starting by K_I or SIF and K_{IC} or fracture toughness followed by G_I, energy release rate and in chapter 9, we described δ which is known by CTOD. It should be remembered that determination of K_I,K_{IC} and δ is possible by analytical-numerical and also experimental method.

However, determination of G_I is possible only for simple cases as we describe by an example herein. The question rose in the past that "is it possible to determine G_I by some approximate numerical method?." The answer was provided by J. Rice in 1968 and he invented a method which is named "*J* integral." This integral sometimes can be determined by analytical method and in general can be computed by numerical methods like FEM.

In this section we demonstrate that as far as LEFM and EPFM is concerned G_I and J either are identical or closely related to each other. Before, that we determine G_I for a simple case by considering a double cantilever beam with a surface crack by length a as shown in Fig. 10.7, the widths in B and thickness is $2h$ and pulled from both sides by pair of concentrated force P as shown in the Fig. 10.7.

Considering the x and y axis in the direction shown by Fig. 10.7, when the crack opens by the force P at $x=0$ the bending moment produced

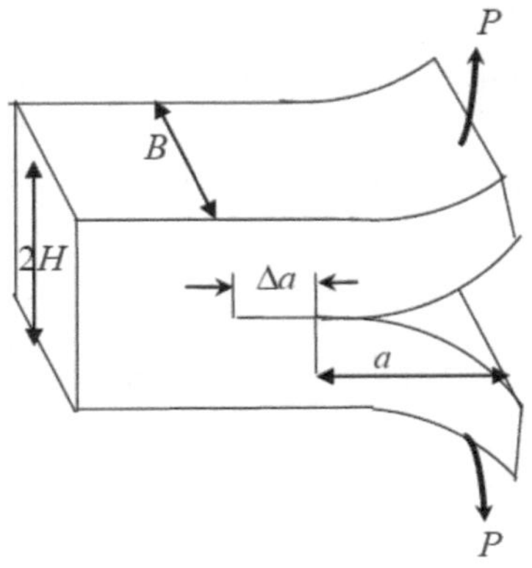

FIGURE 10.7 Double cantilever beam.

in each piece of the beam with length a is a cross section located at x is $M(x) = Px$, the total strain energy of the two pieces is:

$$U = 2\int_0^a \frac{M^2(x)}{2EI}\,dx \tag{10.3.1}$$

$$I = B\frac{h^3}{12} \tag{10.3.2}$$

By substituting Eq. (10.3.2) in Eq. (10.3.1) and integrating, we have:

$$U = 2\int_0^a \frac{P^2x^2}{2EB\,h^3/12}\,dx = \frac{12P^2}{EBh^3}\left(\frac{x^3}{3}\right)_0^a = \frac{4P^2a^3}{EBh^3} \tag{10.3.3}$$

However, in Chapter 7 we saw that the energy release rate is $G_I = \frac{\partial U}{\partial A} = \frac{\partial U}{B\partial a}$ which results:

$$G_I = \frac{\partial}{B\partial a}\left(\frac{4P^2a^3}{EBh^3}\right) = \frac{12P^2a^2}{EB^2h^3} \tag{10.3.4}$$

We start describing J integral by considering the contour Γ with the x and y axis as indicated in Fig. 10.8. The contour Γ is surrounded the crack and the function W or strain energy is defined over the surface by:

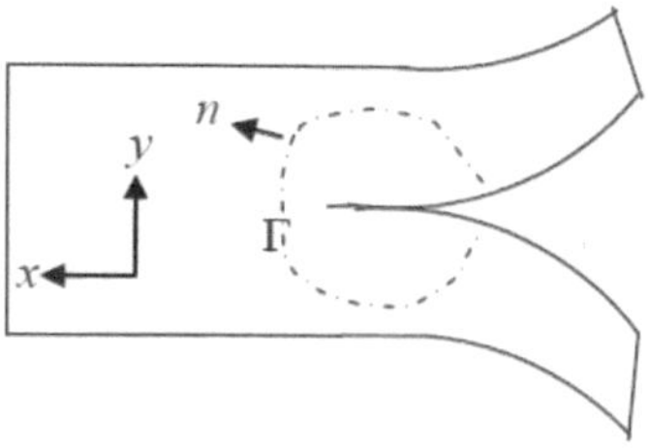

FIGURE 10.8 Integration contour Γ.

$$W = W(x, y) = \int_{0}^{\varepsilon} \sigma_{ij} d\varepsilon_{ij} \tag{10.3.5}$$

For two-dimensional case $i, j = 1,2$ and the traction vector $\vec{T}$, and vector $\vec{n}$ normal to the contour Γ can be defined by the following formulas:

$$\vec{T} = T_x \vec{i} + T_y \vec{j} \tag{10.3.6}$$

$$\vec{n} = n_x \vec{i} + n_y \vec{j} \qquad T_x = \sigma_{xx} n_x + \sigma_{xy} n_y \qquad T_y = \sigma_{yx} n_x + \sigma_{yy} n_y \tag{10.3.7}$$

Based on Eqs. (10.3.5), (10.3.6) and (10.3.7) J. Rice defined the following integral:

$$J = \int_{\Gamma} \left(W\, dy - \vec{T} \cdot \frac{\partial \vec{u}}{\partial x} ds \right) \tag{10.3.8}$$

In Eq. (10.3.8) vector $\vec{u}$ is called displacement vector and can be defined by:

$$\vec{u} = u_x \vec{i} + u_y \vec{j} \tag{10.3.9}$$

In the first instance the J integral seems strange, particularly the term $W\,dy$, but it is obvious that it can be calculated numerically for example by FEM.

To discover the properties of the integral in Eq. (10.3.8), consider a closed contour Γ_c, this counter does not embrace the crack tip (see Fig. 10.9) and designated by $\Gamma_c = \overbrace{ABCDEFA}^{\rightarrow}$ and it contains the following segments:

$$\Gamma_c = \overbrace{ABC}^{\rightarrow} + \overbrace{CD}^{\rightarrow} + \overbrace{DEF}^{\rightarrow} + \overbrace{FA}^{\leftarrow} = \Gamma_1 + \overbrace{CD}^{\rightarrow} + \Gamma_2 + \overbrace{FA}^{\leftarrow}$$

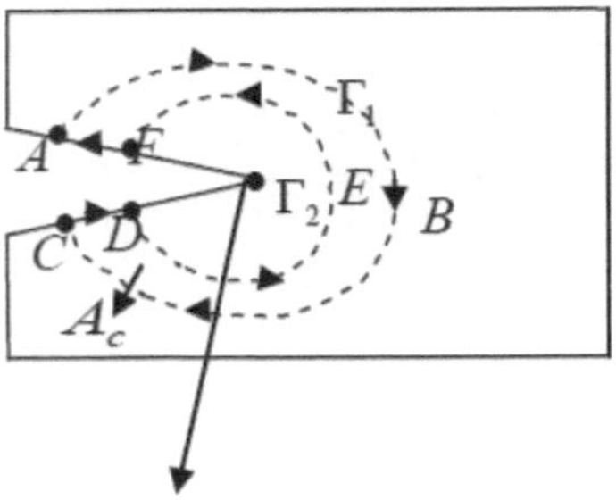

FIGURE 10.9 Closed contour Γ_c.

The above closed contour circles the crack tip but does not enclose it, also in the faces $\overleftarrow{FA}$ and $\overrightarrow{CD}$ there is not traction forces, this means assuming that crack faces should not be loaded. Therefore we can conclude that the function $W\,dy - \vec{T}\cdot\frac{\partial \vec{u}}{\partial x}$ inside the closed contour Γ_c is analytic and in order to perform the integral over the closed contour Γ_c we use Green theorem to convert the line integral to surface integral according to following expression:

$$\int_{\Gamma_c}\left(W\,dy - \vec{T}\cdot\frac{\partial \vec{u}}{\partial x}ds\right) = \iint_{A_c}\left(\frac{\partial W}{\partial x} - \frac{\partial}{\partial x}\left(\vec{T}\cdot\frac{\partial \vec{u}}{\partial x}\right)\right)dx\,dy \qquad (10.3.10)$$

In order to calculate the integral (10.3.10) over the area A_c individual terms in the bracket should be evaluated from Eq. (10.3.5) we have:

$$\frac{\partial W}{\partial x} = \frac{\partial W}{\partial \varepsilon_{ij}}\frac{\partial \varepsilon_{ij}}{\partial x} = \sigma_{ij}\frac{\partial \varepsilon_{ij}}{\partial x} \qquad (10.3.11)$$

But in small deformation we have:

$$\varepsilon_{ij} = \frac{1}{2}\left(\frac{\partial u_i}{\partial x_j} + \frac{\partial u_j}{\partial x_i}\right) \qquad (10.3.12)$$

and if we substitute in Eq. (10.3.11) then:

$$\frac{\partial W}{\partial x} = \frac{1}{2}\sigma_{ij}\frac{\partial \varepsilon_{ij}}{\partial x} = \frac{1}{2}\sigma_{ij}\left(\frac{\partial}{\partial x}\left(\frac{\partial u_i}{\partial x_j}\right) + \frac{\partial}{\partial x}\left(\frac{\partial u_j}{\partial x_i}\right)\right) = \sigma_{ij}\frac{\partial}{\partial x_j}\left(\frac{\partial u_i}{\partial x}\right) \qquad (10.3.13)$$

Final rearrangement in Eq. (10.3.13) is based on the fact that, $\sigma_{ij} = \sigma_{ji}$ and the order of variables x and x_j can be replaced. By considering the expressions in Eq. (10.3.7) we can write:

$$\frac{\partial}{\partial x}\left(\vec{T}\cdot\frac{\partial \vec{u}}{\partial x}\right)=\frac{\partial}{\partial x}\left(\sigma_{ij}\frac{\partial u_i}{\partial x_j}\right)=\sigma_{ij}\frac{\partial u_i}{\partial x_j}\left(\frac{\partial}{\partial x}\frac{\partial u_i}{\partial x_j}\right)+\frac{\partial u_i}{\partial x}\frac{\partial \sigma_{ij}}{\partial x_j} \tag{10.3.14}$$

In Chapter 6, we discussed that for static equilibrium and then we can conclude that:

$$\frac{\partial}{\partial x}\left(\vec{T}\cdot\frac{\partial \vec{u}}{\partial x}\right)=\frac{\partial}{\partial x}\left(\sigma_{ij}\frac{\partial u_i}{\partial x_j}\right)=\sigma_{ij}\frac{\partial u_i}{\partial x_j}\left(\frac{\partial}{\partial x}\frac{\partial u_i}{\partial x_j}\right) \tag{10.3.15}$$

Substituting Eqs. (10.3.13) and (10.3.15) into Eq. (10.3.1) the bracket vanishes and we have:

$$\int_{\Gamma_c}\left(W\,dy-\vec{T}\cdot\frac{\partial \vec{u}}{\partial x}ds\right)=0 \tag{10.3.16}$$

From Eq. (10.3.16) we know that J integral over a closed contour Γ_c is zero. We expand Eq. (10.3.16) into four parts as follows:

$$\int_{\Gamma_1(ABC)}\left(W\,dy-\vec{T}\cdot\frac{\partial \vec{u}}{\partial x}ds\right)+\underbrace{\int_{CD}\left(W\,dy-\vec{T}\cdot\frac{\partial \vec{u}}{\partial x}ds\right)}_{0}+\int_{\Gamma_2(DEF)}\left(W\,dy-\vec{T}\cdot\frac{\partial \vec{u}}{\partial x}ds\right)$$

$$+\underbrace{\int_{FA}\left(W dy-\vec{T}\cdot\frac{\partial \vec{u}}{\partial x}ds\right)}_{0}=0$$ Since the crack faces CD and FA are not loaded, the integral over them vanishes and therefore above equation changes to:

$$\int_{\Gamma_1}\left(W\,dy-\vec{T}\cdot\frac{\partial \vec{u}}{\partial x}ds\right)=-\int_{\Gamma_2}\left(W\,dy-\vec{T}\cdot\frac{\partial \vec{u}}{\partial x}ds\right) \tag{10.3.17}$$

Γ_1 and Γ_2 are arbitrary open contours that circling around the crack tip and minus sign in Eq. (10.3.17) appears only due the opposite direction

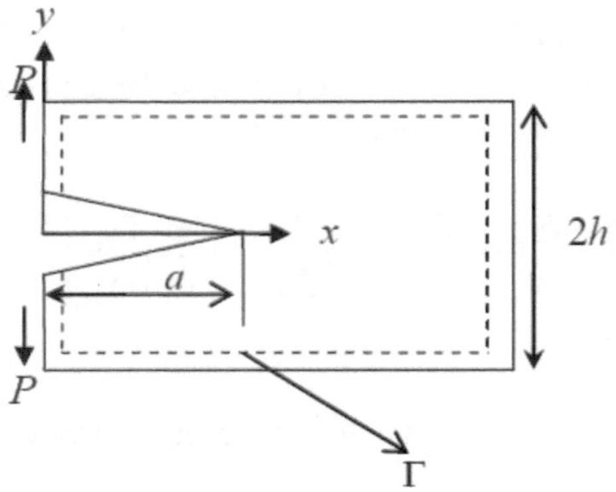

FIGURE 10.10 Contour Γ for cantilever.

(see, Fig. 10.10). From Eq. (10.3.17) we can conclude that J integral over an open contour Γ circling crack tip is not zero and its value is independent of the contour path. Therefore we can choose simple path for a contour to calculate the J integral much easily.

For example if we want to calculate the J integral for the double cantilever shown in Fig. 10.7, according to Fig. 10.10, we choose the contour Γ coincide with edge faces of the cantilever. The traction in the front faces could be found by shear stress results from application of P which is $\tau = \dfrac{P}{B\,h}$ and traction components are $T_y = \dfrac{P}{B\,h}$ and $T_x = 0$, then the individual components of the J integral are:

$$\vec{T} \cdot \frac{\partial \vec{u}}{\partial x} = \left(T_x \vec{i} + T_y \vec{j}\right) \cdot \left(\frac{\partial u_x}{\partial x}\vec{i} + \frac{\partial u_y}{\partial x}\vec{j}\right) = T_y \frac{\partial u_y}{\partial x}$$

Since the boundary Γ, only contain traction $\int_\Gamma (W\,dy) = 0$, then J integral is:

$$J = \int_\Gamma \left(W\,dy - \vec{T} \cdot \frac{\partial \vec{u}}{\partial x} ds\right) = T_y \frac{\partial u_y}{\partial x} h = 2\frac{P}{B}\frac{\partial u_y}{\partial x} \qquad (10.3.18)$$

Now we need to find out about $\dfrac{\partial u_y}{\partial x}$ at $x = 0$, in solid mechanics it is called slope $\theta = \dfrac{\partial u_y}{\partial x}$ or rotation of cantilever with length a, and can be calculated from tables, that is,

$$\theta = \frac{\partial u_y}{\partial x} = \frac{Pa^2}{2EI} = \frac{Pa^2}{2EB h^3/12} = \frac{6Pa^2}{EBh^3} \tag{10.3.19}$$

Substitution of Eq. (10.3.19) into Eq. (10.3.18) results:

$$J = 2\frac{P}{B}\frac{\partial u_y}{\partial x} = 2\frac{P}{B}\frac{6Pa^2}{EBh^3} = \frac{12P^2a^2}{EB^2h^3} \tag{10.3.20}$$

If we compare Eq. (10.3.20) with Eq. (10.3.4), it is obvious that $J = G_I$ and this is valid for LEFM. Therefore J integral which can be computed by FEM, can be replaced by G_I. In EPFM the result may be different and depends on the size of the plastic zone.

In last chapter it was shown that δ_t or CTOD is related to G_I via Eq. (9.4.33), which was, $\delta_t = \frac{G_I}{\sigma_p}$. This section is helped us to demonstrate that as far as LEFM is concerned J integral is also related to G_I (sometimes identical) and the general formula is:

$$\underbrace{J}_{J\,integral} = \underbrace{G_I}_{energy\,release\,rate} = \underbrace{\sigma_p}_{yield\,stress}\ \underbrace{\delta_t}_{CTOD} \tag{10.3.21}$$

In EPFM, Eq. (10.3.21) may not be valid because the applied stress σ is near σ_p and the size of plastic zone increases significantly and the FEM investigations has shown that Eq. (10.3.21) can be valid with a minor modification such that:

$$\underbrace{J}_{J\,integral} = \underbrace{M}_{factor}\ \underbrace{\sigma_p}_{yield\,stress}\ \underbrace{\delta_t}_{CTOD} \tag{10.3.22}$$

Depending on the case the factor M changes within the range $1.15 < M < 2.95$. Equation (10.3.22) is valid only when $\frac{\sigma}{\sigma_p} << 1$ and in higher σ, the parameter G_I is not meaningful and it can be replaced by J.

Further application of J is for the cases when stress strain relation in the material is nonlinear elastic, and it goes up to strain 0.2 level. This is beyond the scope of this section and is the subject of nonlinear fracture mechanics, where the K_I can be expressed by J. The expressions given by Hutchinson are called HRR.

KEYWORD

- **crack face opening**
- **double cantilever beam**
- **elliptical crack**
- **J integral**
- **Westegaard function**

BIBLIOGRAPHY

PART I

1. Green, A. E., Zerna, W. "Theoretical Elasticity", Oxford University Press, 1960.
2. Crisfield, M. A. "Nonlinear Finite Element Analysis of Solids and Structures, Volume 1: Essentials", John Wiley, New York, 1993.
3. Crisfield, M. A. "Nonlinear Finite Element Analysis of Solids and Structures, Volume 2: Advanced Topics", John Wiley, New York, 1997.
4. Simo, J. C., Hughes, T. J. R. "Computational Inelasticity", Springer, 1998.
5. Ogden, R. W. "Elements of the Theory of Finite Elasticity", in Nonlinear Elasticity Theory and Applications, edited by Fu, Y. B. and Ogden, R. W., London Mathematical Society Lecture Note Series, Cambridge University Press, 2001.
6. Belytschko, T., Liu, W. K., Moran, B. "Nonlinear Finite Elements for Continua and Structures", John Wiley, Chichester, 2000.
7. Eringen, A. C. "Nonlinear Theory of Continuous Media", McGraw-Hill, New -York, 1962.
8. Bonet, J., Wood, R. D. "Nonlinear Continuum Mechanics for Finite Element Analysis", Cambridge University Press, 1997.
9. Holzapfel, G. A. "Nonlinear Solid Mechanics: A Continuum Approach for Engineering", Chichester: Wiley, 2000.
10. Sokolnikoff, I. S., "Tensor Analysis: Theory and Applications to Geometry and Mechanics of Continua", Wiley, New York, 1964.
11. Flugge, W., "Tensor Analysis and Continuum Mechanics", Springer-Verlag, Berlin, 1972.

PART II

1. Knott, J. F., " Fundamentals of Fracture Mechanics", London: Butterworths, 1973.
2. Muskhelishvili, N. J., "Some Basic Problems of the Mathematical Theory of Elasticity", Springer; 1977.
3. Parker, P. A., "Mechanics of Fracture and Fatigue: An Introduction", E&FN Spon, May, 1981.
4. Sanford R. J., "Principles of Fracture Mechanics", Prentice Hall; 1st edition, 2002.
5. Zehnder, A. T., "Lecture Notes on Fracture Mechanics", Cornell University, 2007.
6. Irwin, G. R., "Crack Extension Force for Part-Through Crack in a Plate", Journal of Applied Mechanics, Trans. ASME, vol. 29, pp. 651–645, 1962.
7. Rice, J. R., "A Path Independent Integral and the Approximate Analysis of Strain Concentration by Notches and Cracks", Journal of Applied Mechanics, Trans. ASME, vol. 35, pp. 379–386, 1968.

BIBLIOGRAPHY

INDEX

K

L

M

N

O

P

R

S

T

U

V

W

Y

Z